Musings on Light

A Meditative, Non-Mathematical Summary of a Cosmology of Light

Pravir Malik, Ph.D.

ISBN-13: 978-1-7342743-8-7

To The Mother

Author's Introduction

In many respects *Musings on Light* is an image of *A Story of Light*.

A Story of Light is a non-mathematical summary of many of the key concepts in the nine subsequently written Cosmology of Light books that elaborated the mathematical structure and power of light. Through the writing of the nine books, areas that had not been contemplated, and details of the fractal, holographic mathematical structure of light were worked out, to ground and amplify the initial concepts in *A Story of Light*.

By contrast, *Musings on Light* has been written at the other end of the journey, after the completion of all ten books in the two Cosmology of Light book series. It is a contemplative, primarily non-mathematical summary of many of the key concepts that are contained in the ten Cosmology of Light books.

I intended for these musings to be meditative and to assist the reader to intuitively grasp what is meant by Cosmology of Light. To further assist with the meditation, each of the musings of which there are 50, is accompanied by a carefully designed illustration that will further allow the reader to enter more deeply into the body of Cosmology of Light.

Cosmology of Light is intended to be a unified theory of existence from the point of view of light. Hence, it ties together possibility that exists in the native state of light where it is imagined to exist infinitely fast, to the process of multiple quantizations by which light projects itself in layers mirroring symmetrical fourfold realities, to the emergence of space-time-energy-gravity as a cosmic container at the time of the big bang, to the emergence of matter and life in layers of quantum particles, atoms, and cells, to the emergence of humans and culture and civilizations, so that physics, chemistry, biology, psychology, and sociology are all seen as faces of the one body of Light, always behind, immanent in all, and of which all are only richer and richer expressions.

While the first series of six books – *Cosmology of Light*, and the subsequent second series of four books – *Applications in Cosmology of Light*, in totality express an elaborate mathematics for a unified theory of existence from the point of view of light, *Musings on Light,* as a concise summary of the key concepts in the ten books, is by contrast a radically compressed expression of the same unified theory of existence.

It is my belief that developing intimacy with the musings, is the best launching ground for going deeper not only into the two preceding sets of books, but for those so inclined, into the series of concise technical papers extracted from Cosmology of Light, that have been written for IEEE conferences and other journals. An intuitive grasp of the musings will equip the reader with a deeper sense by which to gauge today's surfacing technologies from quantum computation to genetics to artificial intelligence. But even for a reader more inclined to social and human as opposed to technological development, intimacy with the musings will allow a clearer sense for what it may take for collectivities and civilizations to succeed, while also surfacing a more nuanced discrimination for how light-built complexities are involved in any human endeavor.

I would, therefore, recommend *Musings on Light* as being a sound starting point to grasp the body of Cosmology of Light.

Wishing you a light-filled journey of discovery.

Happy Musings!

Pravir,
San Francisco

The Musings

#1: Light's Native State

Light in its native state travels infinitely fast. This is its primal reality.

Due to this infinite speed, it is present across any volume instantaneously. In other words, it is *omnipresent*. If anything, not of the nature of light arises within its fabric it will quickly get overpowered or subsumed by Light. Hence, Light is all-powerful or *omnipotent* within that volume. Since whatever arises or disappears in that volume will do so in its fabric, it will be recorded instantaneously, and therefore Light will have a complete knowledge or *omniscience* in that volume. Finally, since whatever arises will be intimately connected or exist in the one fabric of Light there will be an instant sense of nurturance or harmony in that volume. Hence Light will also be *omninurturing*.

As a species who have evolved in a physical reality where light is known to travel at 186,000 miles per second, or c, in vacuum, our physical senses have been created in and attuned to perceive that reality only. Our entire scientific foundation and all constructed instrumentation is attuned only to this physical reality. Yet the reality of that which cannot be perceived by such narrowly constructed senses, subsequent instrumentation, and subsequent methodology, does not negate its existence.

Let us keep this in mind as we look around us and experience the mystery that we are. There are many implications in considering the fullness of Light and *seeing* what it really is.

These musings from a Cosmology of Light, of which there will be many, will allow us fuller sight into that without which we cannot exist, and yet simply take for granted

#2: The Basis of Existence

Light traveling infinitely fast is the basis of existence. Imagine that everything that arises, does so in this substance of light that is all-present, all-powerful, all-harmonious, and all-knowledgeable. This fourfold omni-reality then, implies too, a basis of infinite information and infinite potential.

As light projects itself at slower and slower speeds, the four properties in light at its native state will exist in all subsequent worlds in some form. And further, these four, being inherently one in light's native state, will be impelled to play together to uphold their deep oneness. Such an embrace of inherent oneness even as the four get separated is the manifest expression of love and implies too, a deep reality of love that exists in light's native state.

Hence, the very basis of all existence is a state of light active with infinite potential, a deep and all-embracing foundation of love, and an increasing and beautiful diversity as the four overarching properties call out more and more of the infinite possibility resident in light.

#3: Quanta and the Big Bang

IN the previous musing we explored the basis of existence as a state where light travels infinitely fast. This state having infinite potential can project itself to simultaneously exist at slower constant speeds. In a Cosmology of Light, when light projects itself at the finite speed of 'c', 186,000 miles per second, this creates the big bang.

But the big bang happens because information in Light traveling faster than speed c seeks to materialize itself through the device of quanta. In other words, quanta are a bridge-mechanism to allow infinite possibility to express itself materially.

Such a rapid materialization of information precipitates a plethora of quantum particles, and this sudden appearance of materialized information is manifested as the big bang.

#4: Light at c is an Intentionality

In the previous musing we explored quanta as the device by which information in Light at its native state, is materialized in slower moving layers of light.

This information is infinitely diverse, embodying as it were, the infinite potential implied by Light's properties of all-presence, all-power, all-knowledge, and all-harmony.

This density of infinite potential can project something of itself at slower speeds of light and in the process create another fourfold reality in which the infinitely subtle, precipitates into some materialized form of itself.

When light projects itself at speed c, this is an intentionality to engender a particular type of materialization and a particular type of reality: there really is no need for light to project itself at speed c, unless there is a particular kind of reality that would be engendered as a result.

Light at c creates a medium as it were, in which a material adventure or manifestation can take place. Light at c is the constant, the root and fountainhead of many material constants, that allows life and further complexity to emerge.

#5: The Emergence of Time and Matter

We know it takes approximately eight minutes for light from the sun to reach earth. Imagine you are on a ray of light and half-way through that journey. As you look back you realize that four minutes ago you were at the sun. as you look forward you realize that in four minutes you are going to be at the earth.

Hence, the experience of past, present, and future arise because of the way light is traveling. By contrast if light were traveling infinitely fast, as it is proposed it is in its native state, then everything would be experienced in the Now.

So, three fundamental properties of our experience of reality arises due to the way light is traveling. But further, matter too arises because of the way light is traveling. As suggested in the musing on the big bang, this occurs when light is projected at c, and is due to the bridge-quanta precipitating vast amounts of information as matter.

Hence the intentionality of light traveling at c is such that it gives rise to past, present, future, and matter.

This significant light-based quadrality — past-present-future-matter — hence, emerges from Light itself.

6: The Essence of a Light-Based Quadrality

In the previous musing it was suggested that past-present-future-matter is a light-based quadrality that arises due to light traveling at speed c. But this can only be if indeed the four emergent aspects of past, present, future, and matter are essentially reflections, or in essence, the same as the four innate properties of light in its native state — omnipresence, omnipotence, omniscience, and omninurturance.

As light materializes at c, since there is infinite possibility that may express itself, there are many such light-based quadralities that will emerge. Past-present-future-matter is a quadrality that expresses itself as part of an intentionality and this becomes clear when the essential symmetry between omnipresence-omnipotence-omniscience-omninurturance and past-present-future-matter is observed.

For the *past* can be thought of as the result of all the activity present everywhere, the *present* can be thought of as the expression of the most energetic or powerful movement in the moment, the *future* can be thought of as the inevitability of knowledge working itself out, and *matter* can be thought of as the increasing play of larger and larger material collectivities working out an essential harmony.

It is this *mapping of an emergent quadrality* to the essential properties of Light in its native state that bestows upon it the status of a light-based quadrality.

#7: The Basis of Einstein's Theory of Relativity

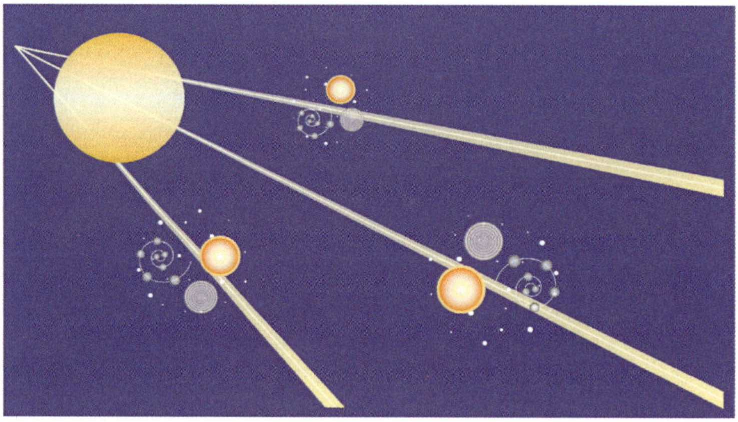

In 'Light at c is an Intentionality' the point made is that light traveling at a constant speed of c has a deeper purpose. This purpose is simply to create a medium in which the infinite potentiality that exists at Light's native state can materialize in a particular way.

It is this realization of light acting as the medium for materialization that is the basis for Einstein's Theory of Relativity. For in relativity, space, time, energy, and gravity are variable and can be made to change depending on the speed and rate of change of speed that an object is traveling at, but the speed of light in vacuum remains constant, regardless.

Further, it can be said that any space-time-energy-gravity composite, whether created through a change in speed, or *ab initio* materializing as such, exists to containerize a wholeness of Light so that some specific functionality in Light's native state can be made to manifest in this layer where light travels at c.

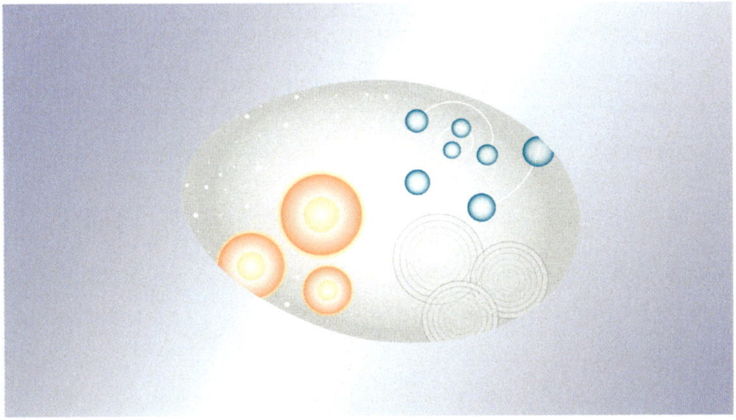

Our experience of past, present, future, and matter arises because of light traveling at c. There is a particular behavior of time and matter that is allowed in the medium of Light existing at c, and the speed of an object existing in this layer of Light cannot therefore exceed c. To transcend that would mean that the 'object' containerized by a specific space-time-energy-gravity composite — which serves the purpose of maturing the specific functionality seeking to materialize — is no longer bound by the laws defined by that space-time-energy-gravity composite, and that would be a contradiction in terms.

But if Light can also exist at different constant speeds as proposed by a Cosmology of Light, then for an object to transcend laws at c becomes possible to the degree to which it opens and anchors in layers of light that exist at speeds greater than c. But this will become easier to the degree that the functionality that needs to materialize within the context of a specific space-time-energy-gravity composite is fulfilled.

This then also becomes the condition by which Einstein's Theory of Relativity might be transcended.

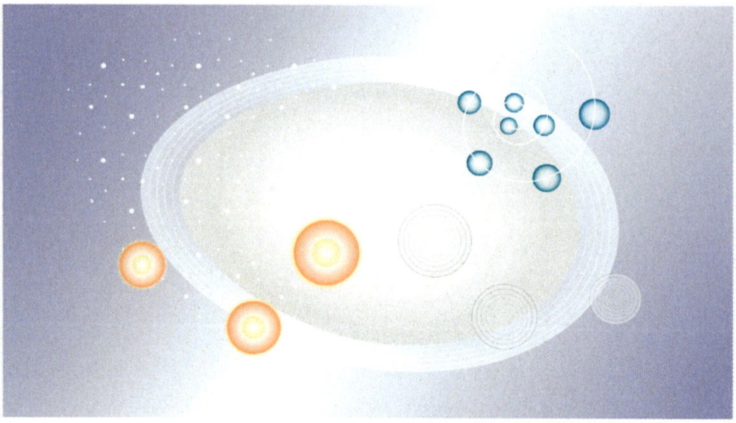

#8: From Unity to Diversity

It is proposed that at Light's native state, where it travels infinitely fast, properties of all-presence, all-power, all-knowledge, and all-harmony are ways in which we may grasp something of its essential unity.

And yet what we experience here in this material world, that has arisen where light travels at c (186,000 miles per second in vacuum), is an immense diversity. No two entities — snowflakes, thumbprints, leaves, persons, ad infinitum — are the same.

So, how does that essential unity become infinite diversity? Mathematically, this would take place via two transformations.

First, where each of the four properties of Presence, Power, Knowledge, and Harmony reveals many shades that are slight variations on the essential property. In other words, Presence becomes a very large set of elements related to 'presence', Knowledge becomes a very large set of elements related to 'knowledge', Power become a very large set of elements related to 'power', and Harmony becomes a very large set of elements related to 'harmony'.

Second, elements from these four sets of practically infinite elements each, combine in unique combinations to create the infinite seeds of diversity.

These seeds in turn materialize through *quanta,* which as suggested in 'Quanta and the big bang', is nothing other than a mechanism to materialize the infinite information that exists in Light.

#9: Bounds of Existence

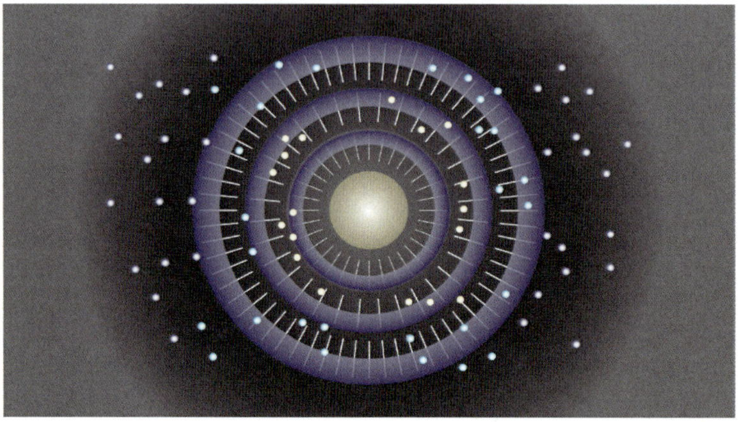

In 'Light at c is an Intentionality' it was suggested that light traveling at *c* creates a necessary medium for reality to emerge as it does, and further that *c* is a fountainhead from which other significant constants also emerge.

When light travels at speed *c*, it means that for it to *express itself or have its light felt* will require it to traverse a "unit" distance. This unit distance is dependent on its speed, and in a Cosmology of Light, must be considered in context to the faster moving layers of light antecedent to this layer in which physical reality emerges.

There is meta-information that is being materialized at light traveling at *c*, and the unit distance required for this expression can be thought as engendering another fundamental constant that is instrumental in bounding existence — Planck's constant, *h*.

The meta-information cannot express itself in halves or fractions but becomes meaningful only as "wholes". Such wholes are 'quanta', or in the case of light 'photons', and light at *c* is so, such that *wholes* can meaningfully materialize. The energy of a photon has in fact formally been formulated to relate to Planck's constant, *h* (re: Energy of photon = h x c / wavelength of light).

But *h* is anchored in *c*, and if the speed of light — *c* — were to change, meaning that the dynamics of what needs to be materialized is changed, then *h* would change as well.

Light at *c*, therefore, not only defines an upper bound to physical existence, but through the creation of *h*, the lower bounds as well.

#10: The Materialization of Light

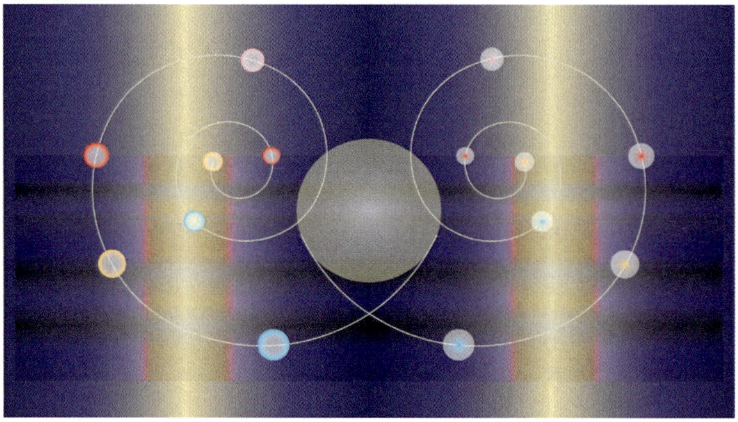

'From Unity to Diversity' suggested that there is a process of mathematical transformations by which the unity of light in its native state becomes the vast diversity, at light traveling at c.

In the first transformation each of Light's essential properties of Presence, Power, Knowledge, and Harmony begins to reveal variations in the essential property through a process of self-reflection. With each reflection a new though similar aspect of the essential property is revealed.

But such a transformation must also have an accompanying or initiating dynamic that results in such a fractal augmentation of the essential properties of Light. In a Cosmology of Light such a dynamic is proposed to be a slowing down of light relative to its infinite speed in its native state.

When light slows down, then the binding factor of matter, *h*, as suggested in 'Bounds of Existence', can change. For, Planck's constant — *h* — in a Cosmology of Light, is a result of the speed of light. Hence, at light speeds greater than *c*, the resulting '*h*' would become smaller than *h* since there is an inverse proportionality between *c* and *h*.

h in this physical realm is such that matter forms in the way that it does. At faster than *c* speeds of light, this binding factor, *h*, would be smaller in value, and hence result in a materialization where matter would not be bound in the way it is when light travels at *c*.

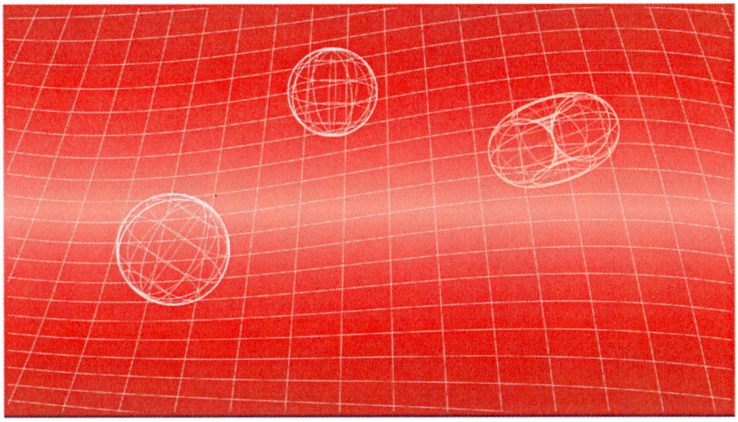

In Light's native state, since the speed of light is infinite, '*h*' would be zero, implying that the essential properties of Light will pervade existence. As light slows down, to create vast sets of properties anchored in each of the four essential properties of Light, '*h*' must be such that it allows the essential properties to splinter into a vast number of variations of itself. Further, '*h*' must be such that it allows these variations to be accessed in any subsequent layer of Light. This may suggest that each element could exist in a relatively materialized '*field-like*' form.

With the second mathematical transformation, light would be slower than in the first transformation, and yet faster than *c*. The resulting '*h*' would hence be such that the material form would be something between being 'field-like' and 'particle-like', the latter being the basis of matter as it begins to manifest in the physical realm where light travels at *c*.

Perhaps such an *'h'* that accompanies the second transformation will allow form to materialize to be *'wave-like'*, which would also have the needed characteristic of linking various elements from their individual 'fields', to create a *unique though subtle wave-form seed.*

It is such a unique wave-form seed that with the existence of the *h* corresponding to *c*, then becomes the basis of the quantum-bridge that allows matter in its known form to arise.

#11: Block-Field-Wave-Particle Quadrality

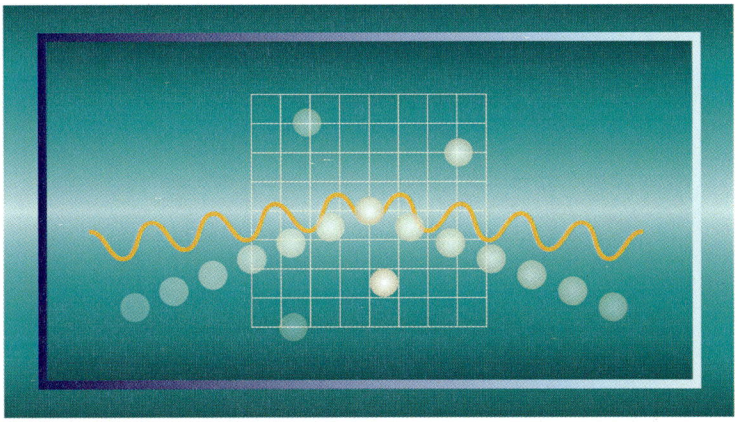

Wave-particle duality is an established phenomenon associated with light and quantum objects. It gets to the idea that neither wave nor particle are sufficient in describing quantum-level phenomenon. While this duality has itself been interpreted in many ways, in a Cosmology of Light wave-particle duality is itself only a partial picture, and *block-field-wave-particle quadrality* provides a more complete view into Light and necessarily into all quantum phenomena that are by definition derivative from Light.

'From Unity to Diversity' provided a high-level overview of key mathematical transformations by which the unity in Light at its native state precipitates into vast diversity at light traveling at *c*. 'Bounds of Existence' introduced the notion of *c* being an upper-bound, and its derivative, *h* — known as Planck's constant and functioning as a binding factor of matter — being a lower-bound in this physical reality where light travels at *c*. 'The Materialization of Light' combined the ideas of the mathematical transformations and the interplay between '*c*' and '*h*', to summarize four distinct forms that Light manifests as in the process of materializing some of the infinite information within it.

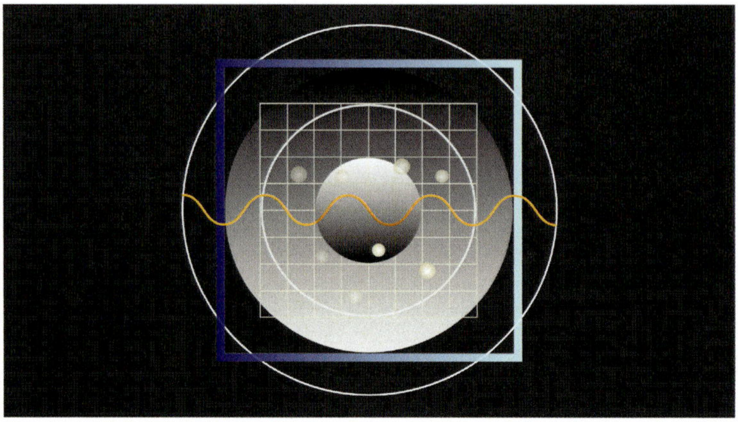

Hence, Light in its native state exists everywhere and is the macro-container within which all other materialization takes place. This macro-container can be thought of as a block — bounded by 4x4 or quadrality-based dynamics — regardless of the point of view. The speed of light being infinite renders an '*h*' or binding factor of matter as zero. Hence the essential properties of Light exist ubiquitously.

When light slows down, relatively speaking, so that '*h*' begins to take on a minuscule value the essential properties begin to separate into numerous fractal reflections that become fields that cut across all existence.

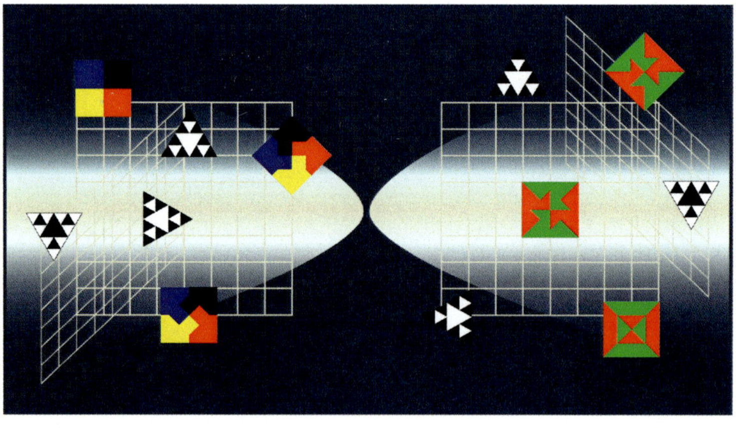

When light slows down even further, so that 'h' increases, but is still smaller than the h associated with c, then 'matter' becomes wave-like, connecting unique elements from the infinite array of fields, and in the process creating a *unique wave-form based seed.*

When light slows down to c, the unique wave-form based seeds begin to particularize through the device of quantum-bridges thereby precipitating into the known bases of physical matter.

Light, and therefore any quantum-object, is in this point of view, a block-field-wave-particle quadrality.

#12: The Logic of Triple-Quantization

The sense of a block-field-wave-particle quadrality might be better grasped when one reflects on the essential mathematical transformations required to take the *input* of unity in Light at its native state, and create the *output* of material diversity when light travels at c.

The first mathematical transformation differentiates the essential properties of Presence, Power, Knowledge, and Harmony into four infinite sets. This kind of materialization is in itself an act of quantization in which information in Light's native state can gather itself in such a way so as to reveal something of itself in the concrete form of field-like sets.

The second mathematical transformation allows formation of seeds derived from unique combinations of elements belonging to the four sets of Light's essential properties. This too is a materialization and can also be perceived as an act of quantization allowing element-based information to gather itself in such a way so as to reveal further concreteness in the form of unique wave-like seeds.

These unique wave-form based seeds then further quantize to become the starting point of physical matter as we know it.

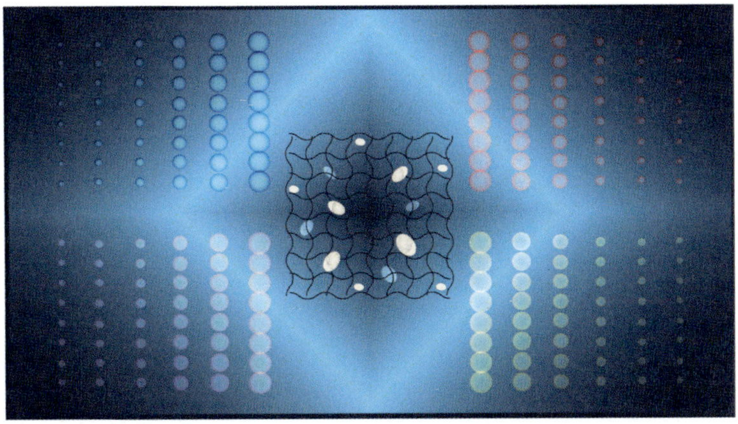

The question that can rise at this point is why would there only be these three types of quantization? The answer is that in this model of cosmology only these three types of quantization, revealing a block-field-wave-particle quadrality in the process, are required for unity at Light's native state to become infinite diversity when light travels at c.

#13: Reframing the Basis of Quantum Computation

In 'The Logic of Triple-Quantization' the three conceived quantizations precipitating something of Light from unity to materialized diversity, are nothing other than instances of quantum computation.

This quantum computation is integral to Light and arbitrates the infinite potential that exists in Light's native state into material reality. This arbitration can be thought of as a persistent cosmic-level quantum computation, recomputing every part of Space, from moment to moment.

This persistent quantum computation suggests the innate creativity natural to such computation. In this creativity something of the infinite potential in Light expresses itself through the device of quanta into more and more material form.

Such quantum computation is innately *creative* and fundamentally differs from the more familiar digital computation that is *constructive*. Such cosmic-level quantum computation also suggests the need for a different paradigm for quantum computers.

The current digitally-inspired constructive focus central to the quantum computing paradigm needs to be reframed to tap into the innate creativity characteristic of the logic of triple-quantization.

#14: The Purpose of Entanglement and Superposition

'Block-Field-Wave-Particle Quadrality' makes the point that Light primarily exists as a block, and can also exist as a field, a wave, and potentially a particle simultaneously.

This simultaneity suggests an alternative interpretation to the quantum phenomena of superposition and entanglement.

By virtue of a block containing all fields, waves, and particles, the fourfold Presence-Power-Knowledge-Harmony reality of Light is intimately entangled in everything. It is such block-type entanglement that ensures that "intelligence" of Light exists across all things.

Similarly, the field-like extension of elements from the four sets suggests a field-type entanglement prior to the formation of seeds of uniqueness. Finally, because of the waveform of seeds, there is an entanglement at the seed level. It is such field-like and wave-like entanglement that ensures infinite diversity of seeds and subsequently the vast diversity at the material-level.

The block-field-wave-particle quadrality also suggests that information from all four layers of light is available to influence the process of materialization through quantum computation. This simultaneity of different layers of information related to a single seed — inputs in the arbitration of material outcomes — is nothing other than superposition.

#15: The Origin and Possibilities of Genetics

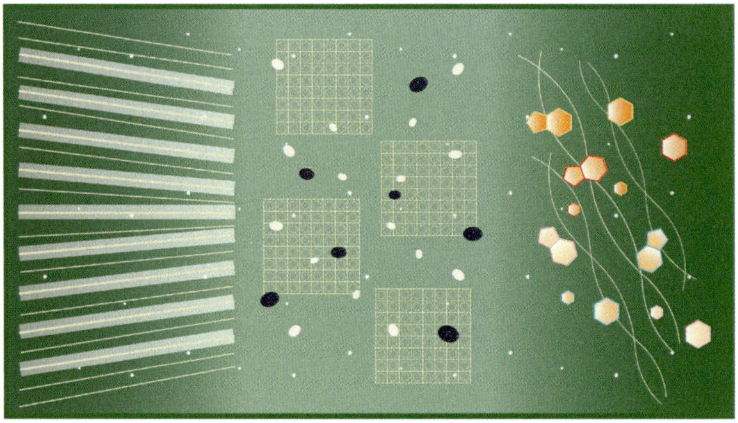

Any process of computation involves inputs, processing, and outputs. The creative basis of quantum computation central to a Cosmology of Light similarly involves inputs, processing, and outputs. The *inputs* derive from information specific to a layer of light, the *processing* leverages triple-quantization, and the *output* is genetic-type information. In a Cosmology of Light this information in faster-moving layers of light is the source of all genetics.

This genetic-type information is different in each layer of Light. In Light's native state the genetic-type information has to do with management of infinite possibility across the block-aspect of Light. In Light's field-aspect the genetic-type information has to do with the four practically infinite-size sets structured around the essential properties of Light. In Light's wave-aspect the information has to do with the potentially infinite unique seeds that drive the materialization of physical matter.

Light's particle-aspect is instrumental in the creation of the material building-blocks that contribute to genes. The quantum arbitration that brings the particles into being relies on the three types of genetic-type information in Light's antecedent block, field, and wave aspects, both overtly and subtly. Overtly, through the direct influence of layer-specific genetic-type information. Subtly, through the process of entanglement and superposition.

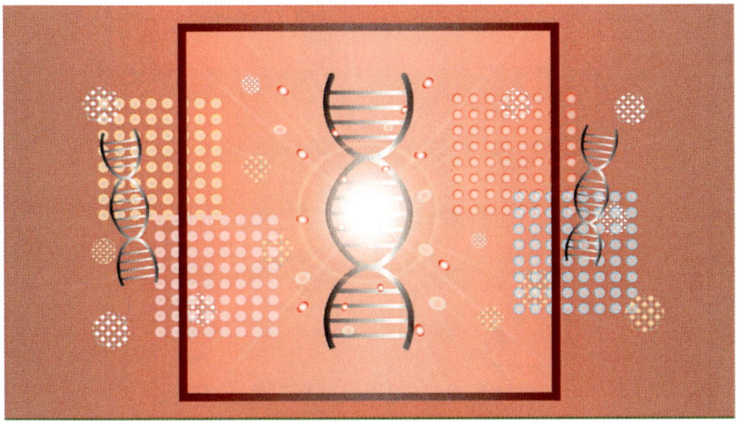

The infinite potential resident in Light's native state, the reality of triple-quantization and persistent cosmic-level quantum computation, implies that while there may be tendencies, nothing, even genes, are ever fixed.

Change can be brought about by invoking Light. The deeper the invocation the more significant will the change likely be.

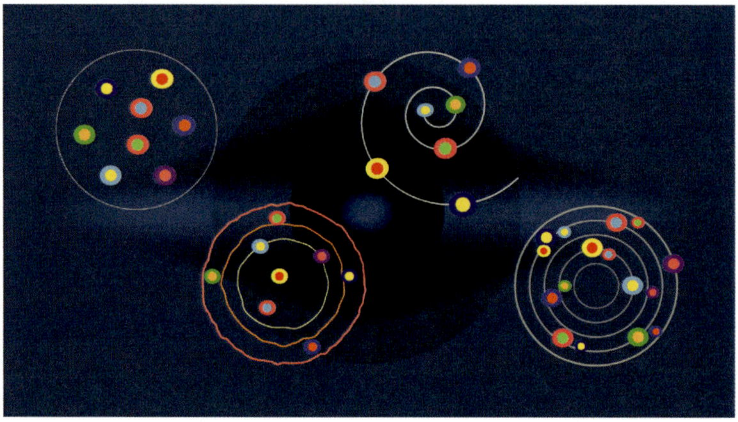

In a Cosmology of Light space, time, energy, and gravity are a light-based quadrality: in other words, they emerge from Light as a cohesive whole. All four of these cosmic parameters are present in the operation of any object or group of objects, regardless of scale.

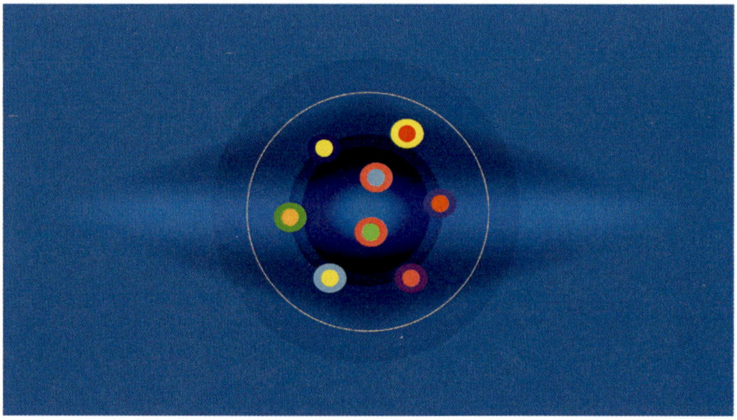

Space can be thought of as pregnant with a large number of seeds. These seeds are the result of a couple of mathematical transformations that essentially allow the infinite unity in Light's native state to become the vast material diversity when light travels at c.

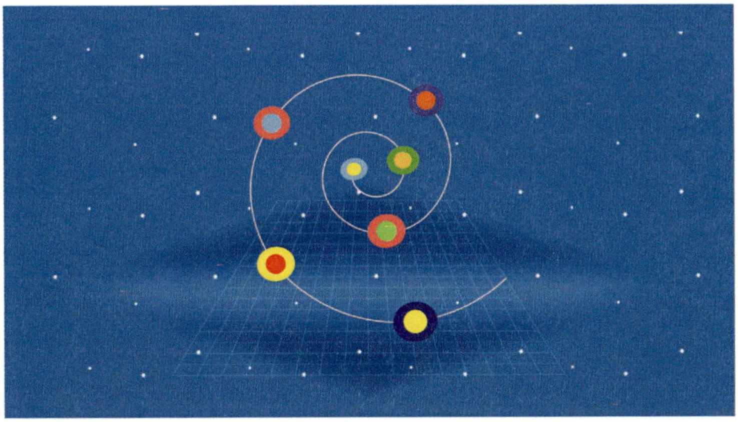

Time can be thought of as the unique path of material maturity of the possibility contained in the seeds in space. Time hence is completely relative and may proceed differently for each seed.

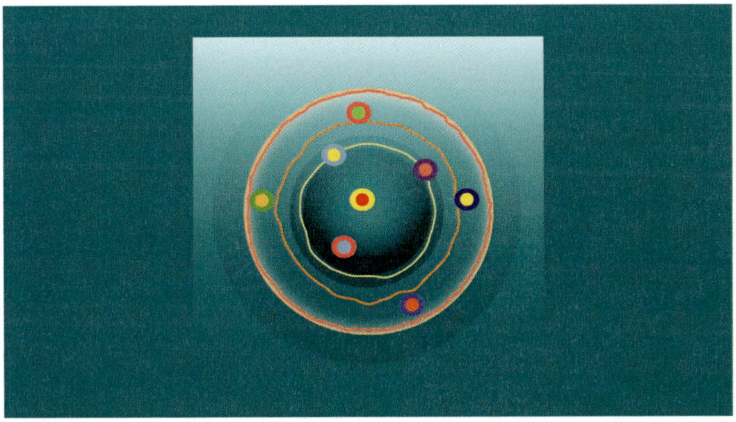

Energy can be thought of as the substance associated with a seed that results in its materialization.

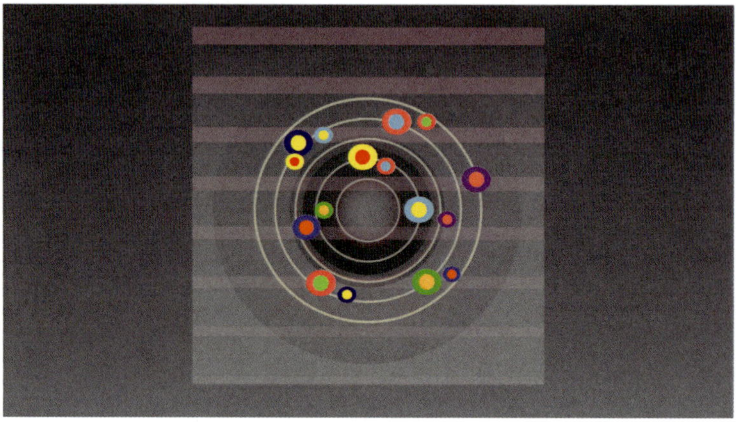

Gravity can be thought of as the arrangement or relationship between seed and seed that allows more and more complex forms of harmony to come into being.

Hence, it is nothing other than the process of the materialization of Light that gives rise to space, time, energy, and gravity.

Space-time-energy-gravity is a light-based quadrality and it can be seen that Space, that holds all archetypes in its bosom, is nothing other than *Light's aspect of Knowledge*. Time, that ensures the material maturity of what is contained in the seed, despite every and all obstacles, is nothing other than *Light's aspect of Power*. Energy, which is the substance that will materialize is nothing other than *Light's aspect of Presence*. Gravity, that allows more and more complete and mature relationships between materialized seed and materialized seed is nothing other than *Light's aspect of Harmony*.

#17: The Fourfold Space-Time-Energy-Gravity Script of Genetics

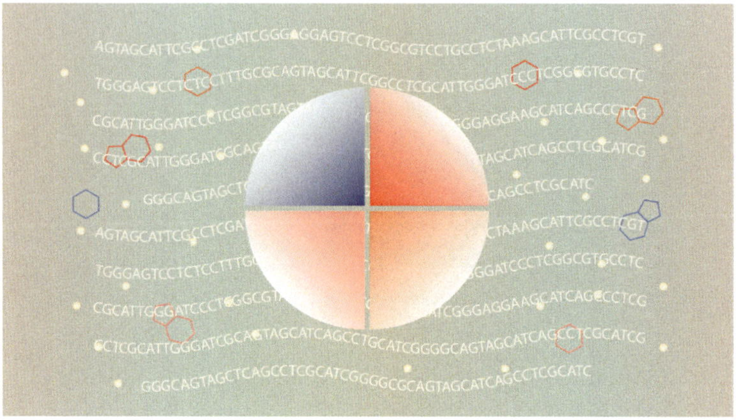

The persistent cosmic-level quantum computation that underlies the Cosmos is a comprehensive arbitration. The input into this arbitration is the variety of possibilities existing in layers of light at different speed. The output is genetic-type code.

As light approaches c, marked by the phenomenon we know as the big bang, then the space-time-energy-gravity quadrality becomes the script of all genetic-type code, whether subtle or material. In other words, genetic-type code, post big bang, is written using a space-time-energy-gravity script.

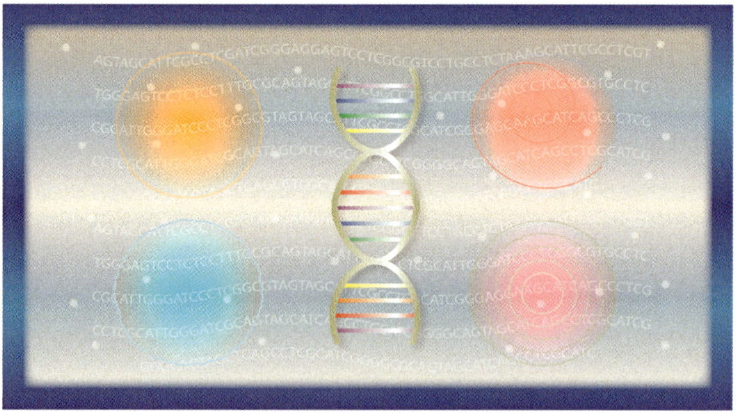

This space-time-energy-gravity script is always written in the medium of light. All such scripts are recorded in the light at the quantum-level veil. Some of these scripts materialize further through quantum-bridges, and at the familiar cellular level this space-time-energy-gravity script materializes or is written in DNA.

#18: Space-Time-Energy-Gravity Macro- and Micro-Containers

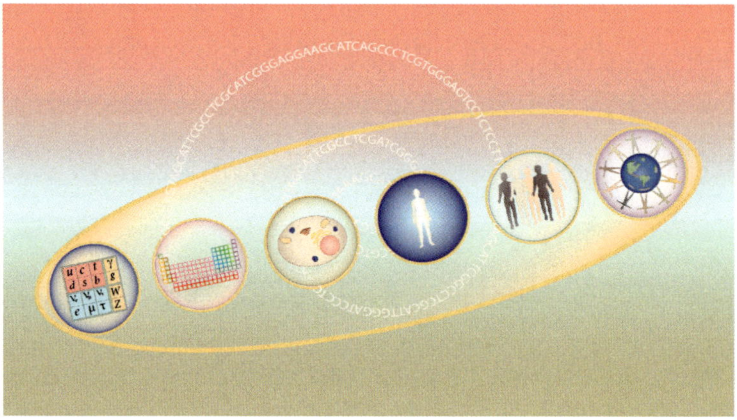

In its path of precipitation as Light approaches or projects itself at speed c, there are key dynamics that have already taken place, involving mathematical transformations, the play of the speed of light 'c' and its inverse Planck's constant 'h', that creates a vast array of 'seeds'.

These 'seeds' as suggested in 'The Mystery of Space, Time, Energy, and Gravity', become Space, Time, Energy, and Gravity. In other words, Space, Time, Energy, and Gravity, are emergent from Light and are the aggregate-level macro-parameters formed by information in deeper layers of light, that create the complex material-field or macro-container within which the further adventure of Light unfolds.

But as suggested in 'The Fourfold Space-Time-Energy-Gravity Script of Genetics', there is a significant micro-level action of space-time-energy-gravity in which detailed dynamics are scripted.

Physical "laws", behaviors of any emergent species, tendencies and biases in more complex organizations ranging from economies to ecosystems, can become the ground for any aspect or variation of the infinite possibility that exists in antecedent layers of Light to materialize, so long as the appropriate space-time-energy-gravity micro-container is altered.

#19: The Truth About the Electromagnetic Spectrum

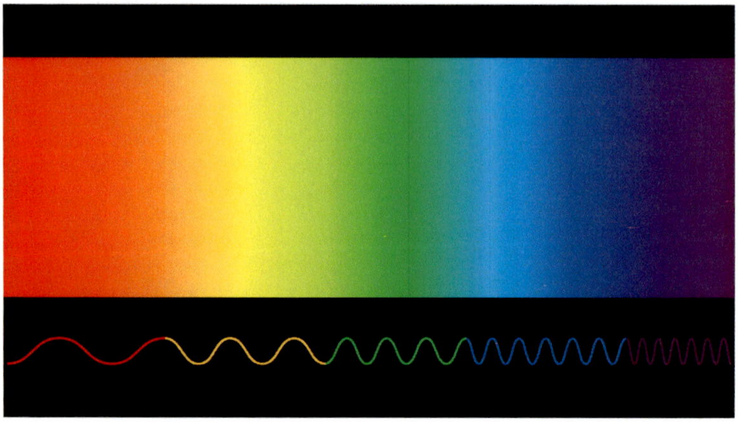

It has been suggested previously that the arbitration of infinite possibility into material reality requires a foundation of light-based quadralities. Yet the common scientific conception of light comprises of an 'electro' and a 'magnetic' aspect only, and the vast range of wavelengths and frequencies gives rise to what has been termed as the Electromagnetic Spectrum.

Ina Cosmology of Light it should be immediately evident that such a description of light is incomplete since any light-based quadrality must have a fourfold presence-power-knowledge-harmony order implicit within it.

While the 'electro' may be equated to 'power', and the 'magnetic' to 'harmony', it is clear that there also needs to be a 'presence' and a 'knowledge' aspect to the "electromagnetic" spectrum.

The range of wavelengths that can be viewed as a collection or array of archetypes of possibility can be seen to comprise the 'knowledge' aspect of the electromagnetic spectrum.

The mass potentials that can arise based on an underlying wavelength, more clearly evident with the rise of organs that can be affected by certain wavelengths of the electromagnetic spectrum, sheds insight into its 'presence' aspect.

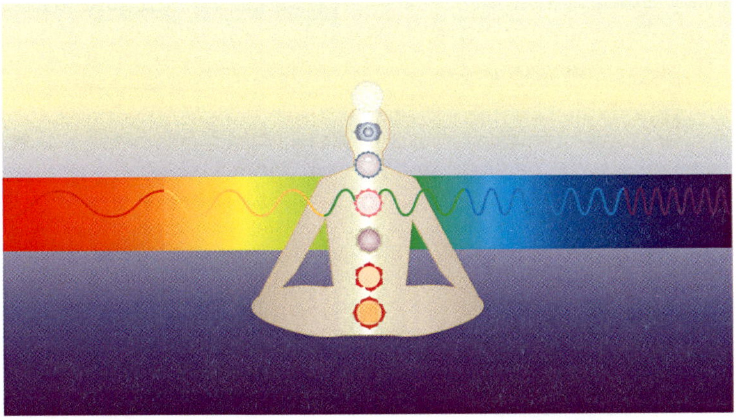

In such a perception the electro-magnetic spectrum can more clearly seen to be an "electro-magnetic-wavearchetype-masspotential" light-based quadrality.

#20: The Quantum Particle Light-Based Quadrality

In a Cosmology of Light, the infinite potential existing at Light's native state requires a series of light-based quadralities to materialize in this realm where light apparently exists at speed c.

Space-time-energy-gravity in both its macro and micro aspects are significant preliminary quadralities. The electro-magnetic-wavearchetype-masspotential spectrum is another. It has already been suggested that quanta are an essential bridge-mechanism and in reality, all quantum particles are themselves a light-based quadrality.

All known quantum particles can be seen to belong to the Standard Model which consists of four types of quantum particles — quarks, leptons, bosons, and the Higgs-boson. Yet when the principle behind each of these types of quantum particles is investigated it can be seen that these are nothing other than a materialization of Light's essential properties of Knowledge, Power, Harmony, and Presence, respectively.

Quarks are the quantum particles that comprise protons. The number of protons in a nucleus of an atom determines its atomic number. The atomic number of an element uniquely characterizes its behavior in the universe. Hence an atom with an atomic number of 47 is Silver and the ways it will behave and interact with other elements is determined by that. In other words, quarks can be thought as the carrier of knowledge, or even as a materialization of Light's aspect of Knowledge.

Electrons as a surrogate for leptons, are strongly associated with energy and power. If an electron shifts orbit in an atom, or is released or captured by an atom, there is a strong display of energy. Hence, leptons may be thought of as a materialization of Light's aspect of Power.

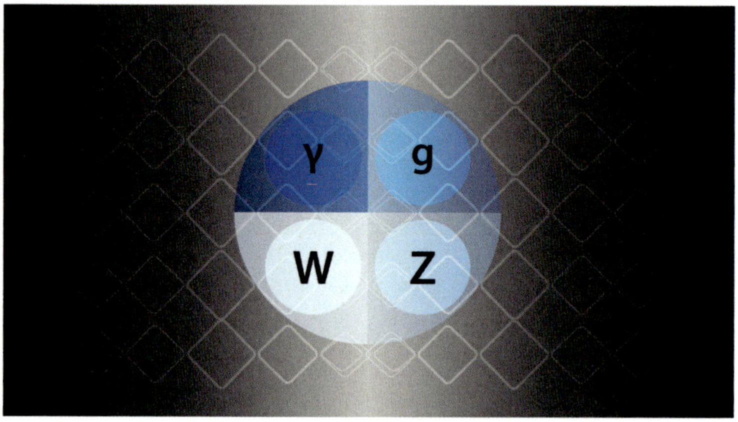

Bosons are force-carriers and essentially allow all other quantum particles to interact and combine. They can hence be thought of as a materialization of Light's aspect of Harmony.

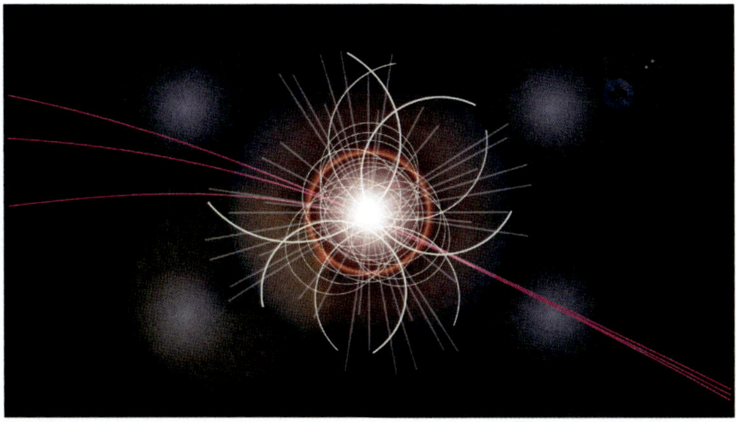

The Higgs-boson allows quantum particles destined to have mass to interact with the Higgs-field to in fact gain mass. This provides material presence to these quantum particles, and hence the Higgs-boson can be thought of as a materialization of Light's aspect of Presence.

Thus arises the quantum particle or the quark-lepton-boson-Higgsboson light-based quadrality.

#21: The Stability of Light-Based Quadralities

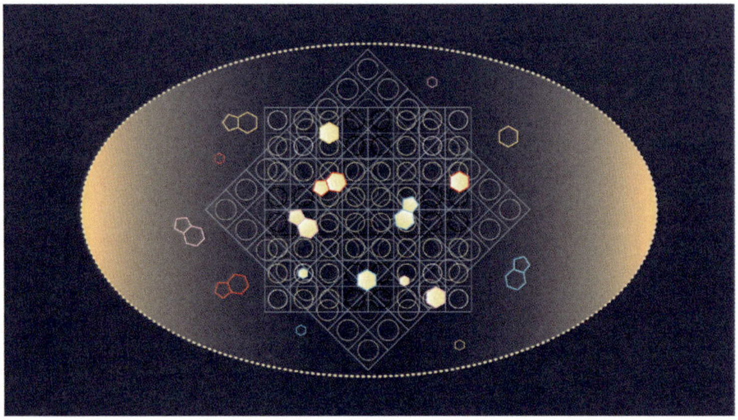

The persistent cosmic-level quantum computation arbitrates reality based on emergences and dynamics already in play at the layer where light travels at c, and possibilities that continue to exist in antecedent layers of light.

The output of such computation is genetic-type information, that as previously suggested becomes "law" by being added to the aggregate space-time-energy-gravity macro-container.

Any light-based quadrality — that by definition marks some milestone in the output of the constant quantum-level computation — is implicitly stable and becomes a foundation for further light-based quadralities to also emerge. Such stability must be, since the implicit unity of properties of light held within itself, seeks to maintain that same unity even as new light-based quadralities emerge in a unique display of possibilities.

Hence it is that light-based quadralities such as the space-time-energy-gravity containers, the electro-magnetic-wavearchetype-masspotential spectrum, quark-lepton-boson-Higgsboson quantum particles, add to the material foundation which forms the growing edifice of Light's continuing materialization of itself.

#22: The Atom Light-Based Quadrality

With the emergence of quantum particles as a stable light-based quadrality, the stage is set for further arbitration of possibilities and the subsequent display of the fourfoldness of light as atoms.

The most common classification of the Periodic Table is in fact fourfold — as s-Shell, p-Shell, d-Shell, and f-Shell atoms respectively.

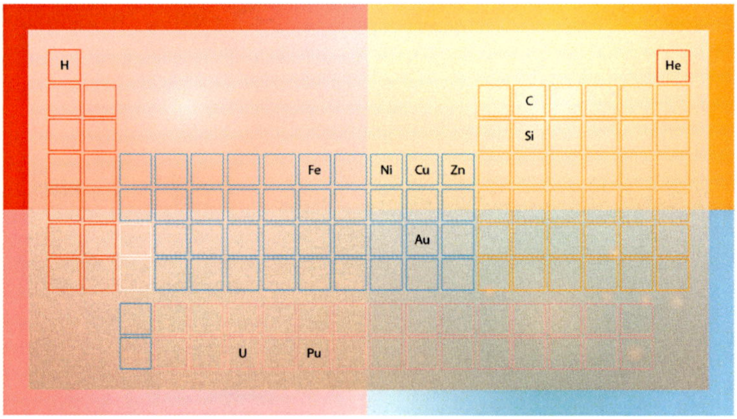

s-Shell atoms, containing the likes of hydrogen and helium, which are instrumental in the process of nuclear fusion that is known to power stars, can be thought of as carriers or a materialization of light's aspect of power.

p-Shell atoms, containing the likes of silicon and carbon, the former being the basis of computing machinery and the latter of complex life-forms holding mind — even if latent — in its crucible, can be thought of as a materialization of light's aspect of knowledge. This notion of p-Shell atoms materializing knowledge is further reinforced by the fact that this grouping also has a number of sub-groupings or archetypes such as metals, metalloids, non-metals, noble gas, and halogens, within it.

d-Shell atoms, containing the likes of iron, nickel, copper, zinc, gold, amongst many others — the 'workhorse' or industrial-level elements — suggests that this grouping is a materialization of light's aspect of presence.

f-shell atoms, those with the largest atomic numbers, and containing both the stable lanthanides and radioactive actinides, appears to be an experimentation in collectivity at the atomic level, and hence can be thought of as a materialization of light's aspect of harmony.

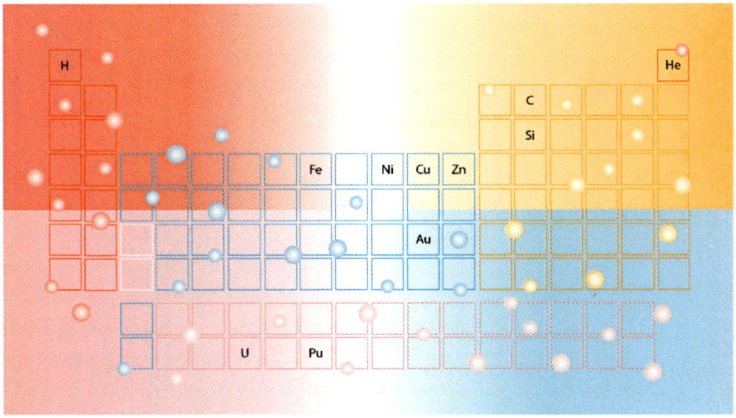

Hence, from a Cosmology of Light point of view, it can be seen that the gamut of atoms as summarized by the Periodic Table not only materializes more of the infinite possibility in light but further, can be entirely classified by the fourfold properties of light hence also giving the collectivity of atoms the status of a light-based quadrality.

The stability of such light-based quadralities will allow the emergence of myriad additional forms of complexity.

#23: The Photon-Gluon-W&ZBoson-Graviton Light-Based Quadrality

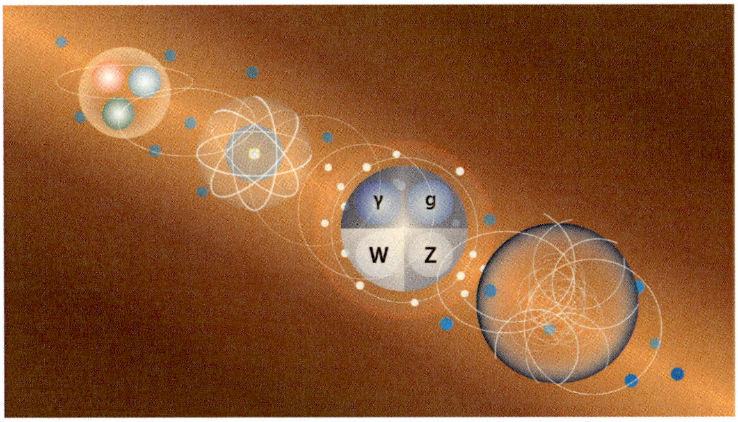

In looking at the quantum-particle light-based quadrality bosons were identified as a materialization of Light's aspect of harmony. It is the action of bosons that allows the vast interplay at the inter- and intra-atomic level, and in this musing, the basis of this will be further explored through a look at the four types of bosons.

Photons are known to be the carrier of the electro-magnetic-wavearchetype-masspotential spectrum. Since this spectrum pervades and effectively connects everything, photons can therefore be thought of as a materialization of Light's aspect of presence.

Gluons are the carrier particle of the strong nuclear force and hold quarks together in their inherently composite arrangements. But it has been posited that quarks are themselves a materialization of Light's aspect of knowledge, and therefore gluons must be related to knowledge, and are also suggested to be a materialization of Light's aspect of knowledge at the quantum force-carrier level.

W and Z bosons are the carrier particle for the weak interactions, responsible for the decay of more massive quarks and leptons into lighter quarks and leptons. This decay is accompanied by the release of power and energy and so it must be that W and Z bosons are an emergence or materialization of Light's aspect of power at the level of quantum force-carriers.

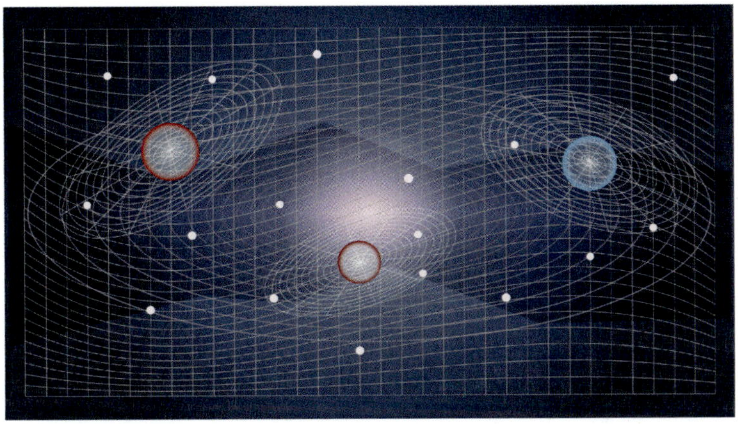

The hypothetical graviton is thought to be the carrier particle for the force of gravity. But gravity holds objects together in relationship. Hence it must be that gravitons are a materialization of Light's property of harmony at the level of quantum force-carriers.

These quantum-level force-carriers then become a photon-gluon-W&Zboson-graviton light-based quadrality.

#24: NucleicAcid-Polysaccharide-Lipid-Protein Light-Based Quadrality

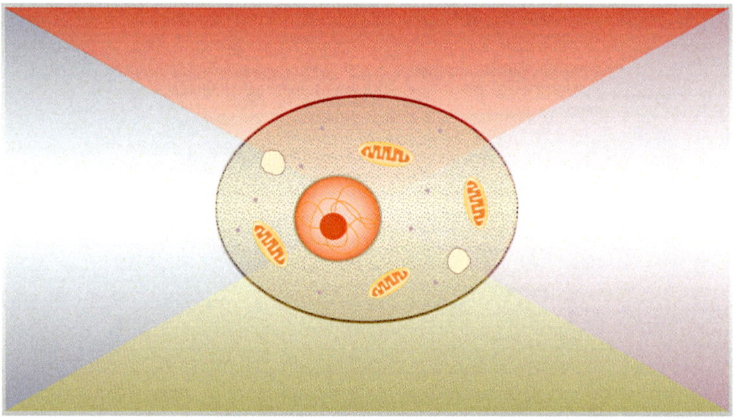

In its continued material expression, the infinite potential in Light, organized as four overarching principles becomes the basis of universal cosmic parameters of space, time, energy, and gravity, the basis of the pre-physical electro-magnetic-wavearchetype-masspotential system, the basis of the physical quark-lepton-boson-Higgsboson quantum-particle system, and the basis of the chemical 's'Shell-'p'Shell-'d'Shell-'f'Shell element system.

In this musing, we reflect on how Light also becomes the basis of all biological systems. Essentially every living cell is animated by four molecular plans. These are, respectively, nucleic acids, polysaccharides, lipids, and proteins.

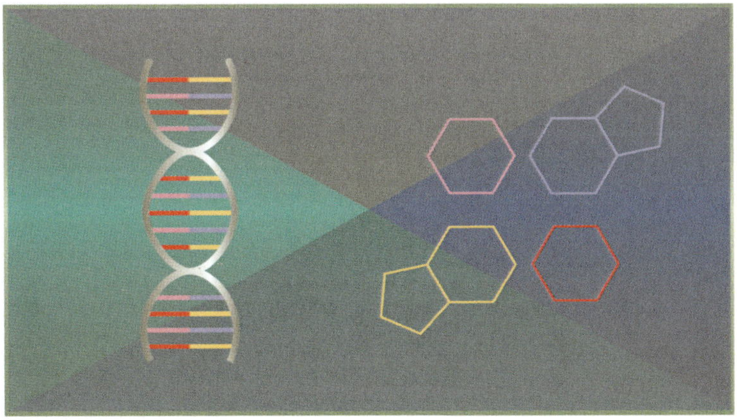

But on examination of nucleic acids, we find that this is none other than a materialization of Light's aspect of knowledge. This is so because nucleic acids create DNA in cells which we know is the repository of knowledge in the cell.

Polysaccharides are simple sugars that are sources of energy and power in the cell. They are, hence, a materialization of Light's aspect of power in the cell.

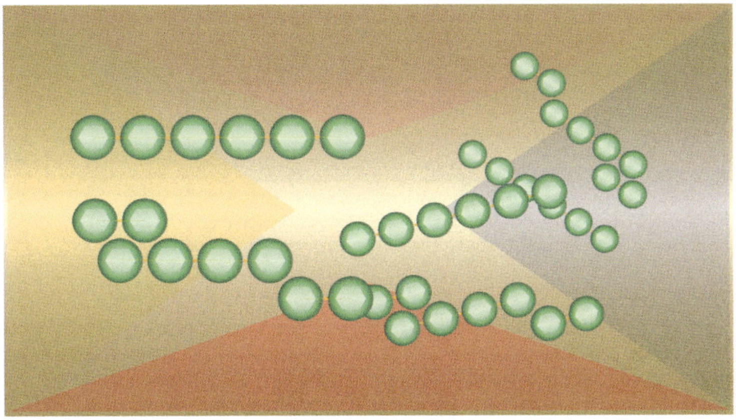

Lipids create compartments in the cell and allow specialization of work. This allows harmony to be established in the cell. Lipids, hence, are a materialization of Light's aspect of harmony in the cell.

Proteins are workhorses. In a human cell, for example, there are over 30,000 proteins doing all the work. These proteins are instruments of service, diligence, perseverance — those aspects that make up the vast activity in a cell hence defining the very notion of presence. Proteins, therefore, are a materialization of Light's aspect of presence in the cell.

These four universal molecular plans, present in all living cells and founded on Light's properties of knowledge, power, presence, and harmony, hence become a nucleicacid-polysaccharide-lipid-protein light-based quadrality.

#25: Expansion and Contraction Dynamics of the Cosmos

In a Cosmology of Light, the Cosmos is essentially a play of Light. Light is the foundation and may exist at different constant speeds. Materialization itself is due to the slowing down of light, so that the infinite possibility in light at its native state, can express itself in unique space-time-energy-gravity-bound containers at light traveling at speed c.

The mystery of space-time-energy-gravity suggests that the primary driver of this composite is in fact unique seeds formed through an antecedent process involving multiple layers of light. The implication of this is that so long as unique seeds or unique functions continue to be created, the space-time-energy-gravity composite will get enriched, the result of which is an expansion of 'Space' or Cosmos.

Enrichment of such functionality is evident when one observes the additional ordered complexity that continues to materialize in the on-going display of a range of light-based quadralities that start from the space-time-energy-gravity quadrality, continues through the electro-magnetic-wavearchetype-masspotential quadrality, the quark-lepton-boson-Higgsboson quadrality, the 's'Shell-'p'Shell-'d'Shell-'f'Shell quadrality, and the nucleicacid-polysaccharide-lipid-protein quadrality.

Such continued materialization implies the existence of more and more unique functional seeds, thus driving the expansion of Space, and the subsequent observed acceleration of galaxies away from one another, in spite of the theoretical pull of gravity that would on the contrary cause the collapse of the universe as its mass increased.

Stable cellular life, the output of the nucleicacid-polysaccharide-lipid-protein quadrality, results in a plethora of living beings. The question then is can such unique functionality continue to materialize with the advent of humans and human-based institutions? In a Cosmology of Light, the very expansion of the Cosmos would depend on this.

Future musings will explore this in more detail.

26: The Human Light-Based Quadrality

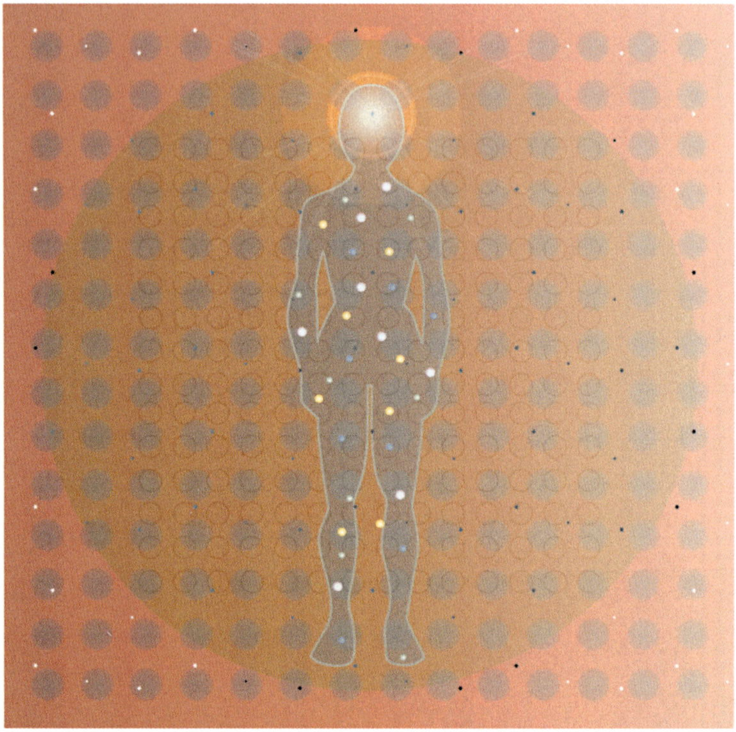

In 'Expansion and Contraction Dynamics of the Universe', it was suggested that the expansion of the universe is going to depend upon the trend of an increase in the materialization of unique functionality: this of course, because the seeding of Space then continues uninterrupted.

There is an evident increase in such materialized functionality as observed in subsequent light-based quadralities starting from the space-time-energy-gravity quadrality, the electro-magnetic-wavearchetype-masspotential quadrality, the quark-lepton-boson-Higgsboson quadrality, the 's'Shell-'p'Shell-'d'Shell-'f'Shell quadrality, and culminating in the nucleicacid-polysaccharide-lipid-protein quadrality.

The first question then, is whether human-beings are also light-based quadralities? For this to be the case the fourfold knowledge-power-harmony-presence properties of Light would have to drive the capacities of a human being.

This is in fact the case…

Thoughts are a play of the emergence of Light's property of Knowledge. Through the thought, we can become greater or conceptualize things greater or begun to enter into relationship with some things other than our small self. Thought, therefore, is at least an instrument in connecting us conceptually to more "othernesses", thereby being none other than a dynamic of knowledge.

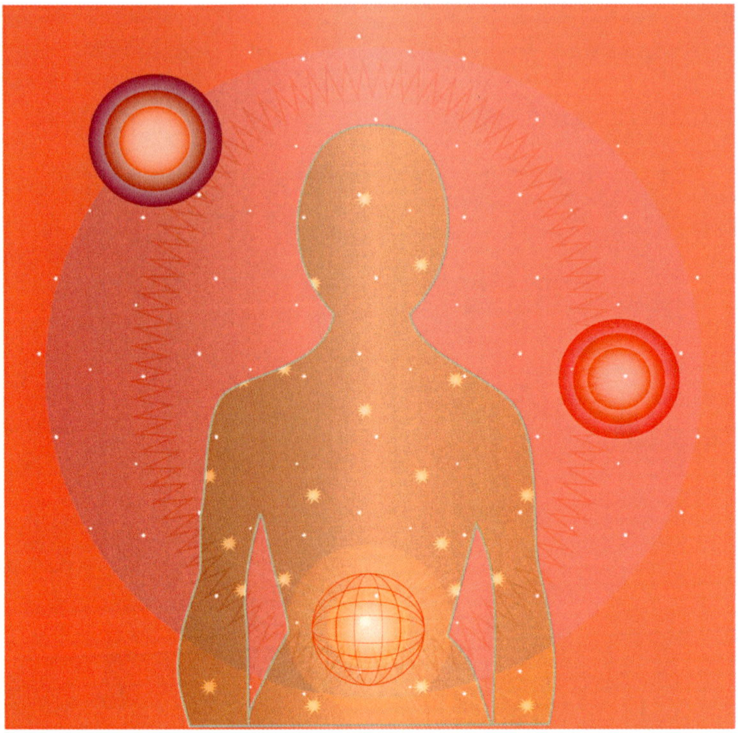

Urges, desires, and wills, are a play of the emergence of Light's property of Power. In the mystery of focus, the vastness of Light has projected itself in us into an apparent smallness that is in reality everything that is. And this smallness is trying through urge and desire and will to connect viscerally or even intentionally to other smallnesses that similarly are nothing other than the fullness of Light projected into a small smorgasbord of selected function.

Feelings and emotions are a play of the emergence of Light's property of Harmony. Its instrument is the Heart, and it generates an array of emotions that are an indication or active radar of whether we are moving toward or away from a reality of harmony, whether based on our small self or some larger Self of Light.

Sensations are those things we experience with our senses. We see things, hear things, and smell things, taste things, and can touch things. This ability to enter into relationship with objects through sensation is nothing other than a result of the emergence of Light's property of Presence. We become present to Presence through the device of sensation.

So, we see that the fourfold properties of Light have seeded our very beings. We are in reality nothing other than something of Light. Whatever else of Light we may be, we are at least a light-based quadrality.

#27: Expansion-Contraction Dynamics of the Human Light-Based Quadrality

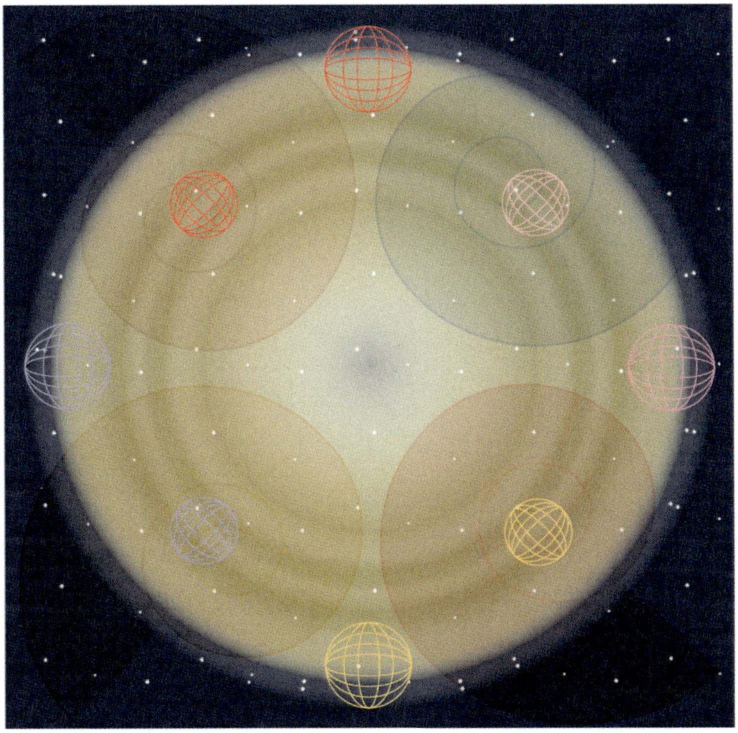

'The Human Light-Based Quadrality' affirmed that humans are indeed light-based quadralities. Humans then must become a means by which the rich fourfold functionality resident in Light, can continue to materialize.

If this did not happen then the compromise to the seeding of unique functionality will not only alter the expansion-contraction dynamics of the universe but also of well-being and expansion at the individual level.

This means that Light's fourfold properties of knowledge-power-presence-harmony displayed as thoughts, will and desires and urges, sensations, and emotions, respectively, must increase rather than decrease their effective dynamism and range of applicability.

With thoughts, are we always trying to calculate, conceive of, and think of how we are only going to fulfill some utterly small desire, or instead do we use this faculty to penetrate more firmly into vaster and vaster mysteries to strengthen relationship with perceived otherness?

With urges, desires, and wills is it just a play of smallness and immediateness that we continually engage in, or instead are we letting something more of Light's infinity of Power approach its fullness through an increasing impersonality?

With our range of sensations are we moving beyond the immediate object-relationship so that the very nature of perception is altered? When we see things, for instance, what are we seeing? Is it just the surface rendering of the play of matter, or do we see that the fullness of Light is still there, with all its potentiality and possibility, in the smallest thing we look at? Do we see that the whole universe and more is present in all its fullness in the least thing that we easily ignore, or belittle, or loathe? When we touch things is it the seeming concreteness of the play of the particles or atoms or chains of molecules that we touch, or is it the Love and Light and the vastness of all that IS that allows itself to be as a small corner that we touch so as to make infinity be felt by something so finite?

With feelings and emotions are we just trying to adhere to a narrow conception of personal harmony, or can we instead actually get to a state where we always feel positive emotions regardless of circumstance? This would mean that we have more truly entered into a relationship with some larger continent of Light.

Our every choice will determine whether we are contracting — by allowing smallness to coalesce, or instead expanding — so as to give Light more room to operate. Dynamics of contraction will precipitate a cycle of disease and destruction of well-being, while the dynamics of expansion will allow wholeness and more of the reality of what Light is to materialize.

#28: On Being, Choice, and Power

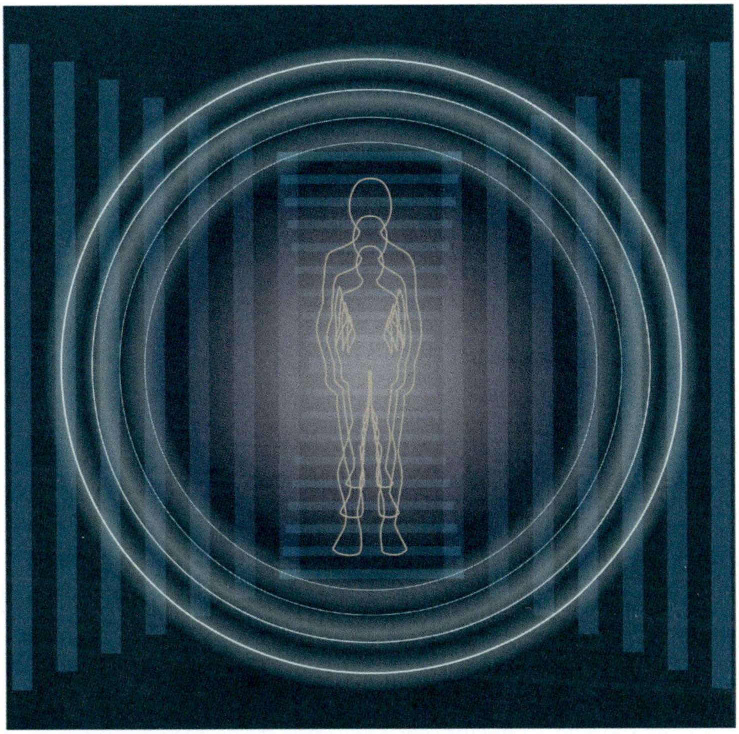

'Expansion-Contraction Dynamics of the Human Light-Based Quadrality' and 'Expansion and Contraction Dynamics of the Cosmos' draw attention to the important dynamic of choice that becomes a critical lever in affecting the micro- and macro-states of well-being and possibility.

But such choice only comes more forcefully to the surface with the emergence of the human being. Prior to this emergence Light projects itself in a series of beings in which decision is as it were, made in the recesses of Light itself.

Hence, beings of a progressively different nature clothe the One reality of Light in forms we refer to as the space-time-energy-gravity quadrality, the electro-magnetic-wavearchetype-masspotential quadrality, the quark-lepton-boson-Higgsboson quadrality, the 's'Shell-'p'Shell-'d'Shell-

'f'Shell quadrality, and the nucleicacid-polysaccharide-lipid-protein quadrality.

With the emergence of the human thought-will-emotion-sensation quadrality, beingness becomes more capable of determining future materialization of the vast potentiality resident in Light. Choice made by the human then becomes critical in how things will evolve. Choice that allows a vaster play of Light in seeming smallness will allow expansion of possibility at both the micro and macro levels.

Increasingly, complex global realities should be seen as a challenge and call to the human-being to establish a richer reality of fourfoldness by allowing more of Light to materialize. All unfolding organizational plays at any level of granularity — family, corporation, village, region, country, economy, civilization — must be seen from this lens. Do we make the choice that shrinks possibility so that organizational plays become dysfunctional and degenerate, or do we make the choice that allows greater largeness and possibility to be established?

That indeed is the power we have been invested with.

#29: The Basis of Complex Adaptive Systems

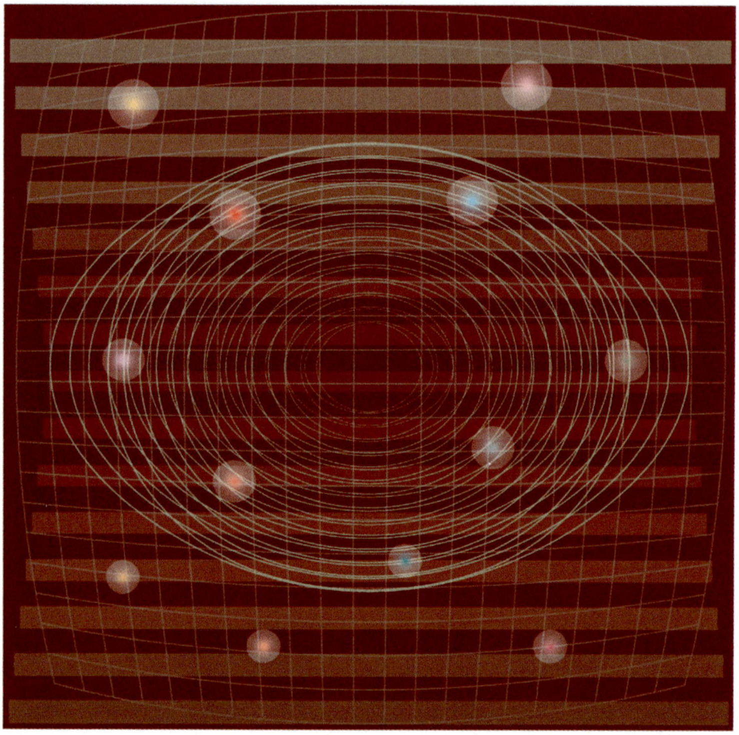

There is a different nature to change that exists in different light-based quadralities.

The space-time-energy-gravity quadralities are fundamental to the layer where light travels at speed c, and both the micro-level space-time-energy-gravity quadrality and the macro-level space-time-energy-gravity quadrality continually change with the ongoing arbitration by which the infinite potential in Light's native state materializes in light at speed c. As a reminder, the micro-level space-time-energy-gravity quadrality is where the detailed arbitration for any emergence takes place, while the macro-level space-time-energy-gravity quadrality aggregates micro-change to enhance the logic of Space, Time, Energy, and Gravity.

Emergences such as the electromagnetic light-based quadrality, the quantum particle light-based quadrality, and the atom light-based quadrality are relatively fixed compared to subsequent emergences, thereby also creating the material foundation that then allows the vastness of the play of molecules and cellular-level molecular plans at the heart of any cell to come into being.

With each subsequent emergence, the layering of material levels adds to the complexity of phenomena and creates the basis of what is known as Complex Adaptive Systems (CAS).

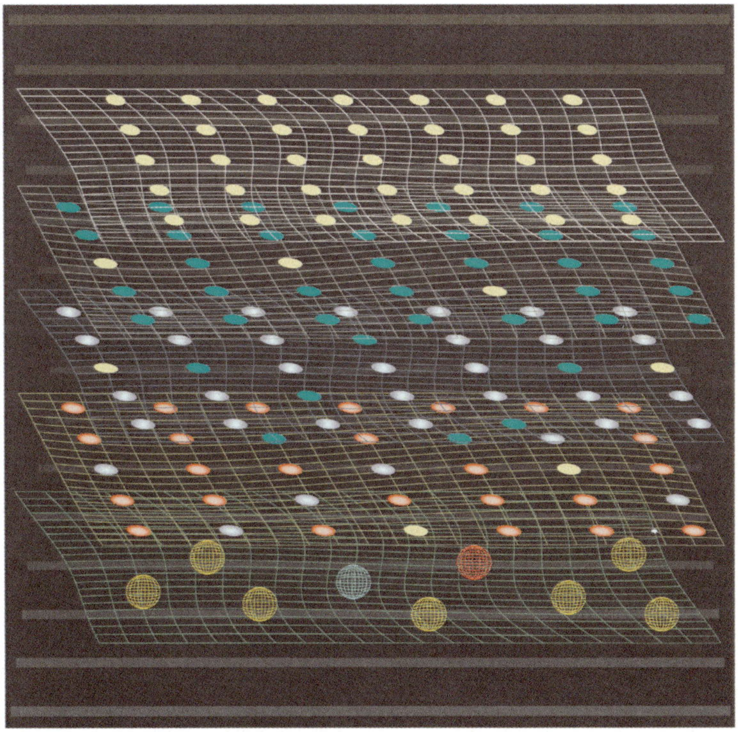

The emergence of such layers that each runs on an independently developed logic — as specified by the associated space-time-energy-gravity micro-container — and that then interact with each other so that a change or agent at one layer can affect a layer at another level of granularity is the cornerstone of any CAS. It is such interaction between independent "agents" both within and between layers that creates feedback loops — with all the paradoxes, uncertainties, and apparent chaos — that CAS are known for.

In 'Being, Choice, and Power' the point made is that the nature of emergent systems changes due to the very dynamic of choice. Hence human CAS are going to be of an entirely different nature — with even more paradox, uncertainty, chaos, and the possibility of unexpected emergence — than any pre-human CAS. Complexity is going to be magnified when we, as humans, lose track of the underlying urge in Light to continue to express or materialize the vast fourfold possibility embedded within it.

Our choices then put us in opposition to the underlying urge to materialize fourfoldness, and this then easily creates a slew of social, environmental, political, and economic crises in its wake. Some of the dynamics and even the necessity of such crises will be examined in future musings.

#30: The Importance of Culture

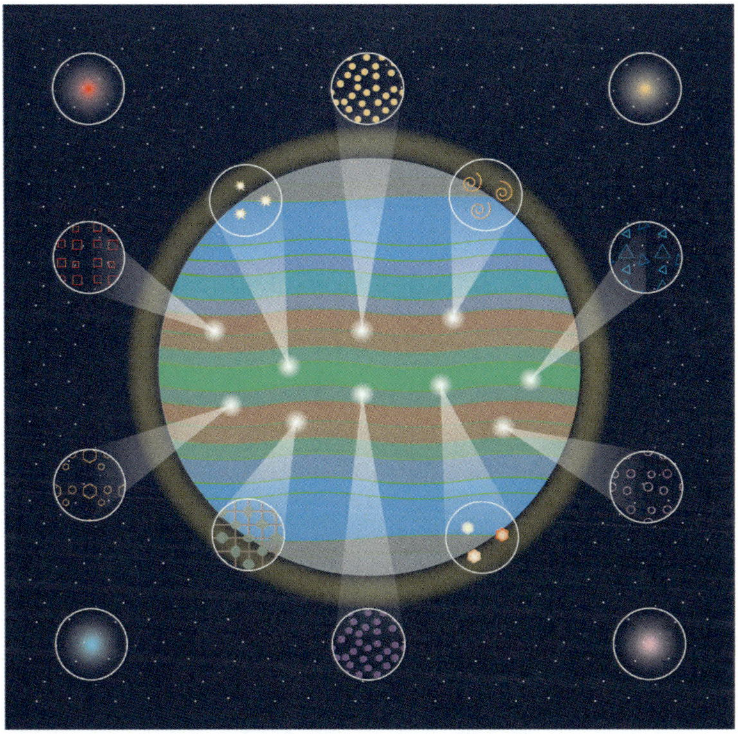

The point of Light traveling at speed c, is to effectuate a massive materialization of the infinity in Light's native state. This materialization will continue so long as Light's fourfold potential continues to express itself. This has been evident in the arc leading from the initial space-time-energy-gravity light-based quadrality to the human light-based quadrality.

It is easy though for a human, endowed with choice, and the complexities of incarnating in an initially narrow thought-will-feeling-sensation container, to reverse the expression of fourfoldness that is the cornerstone of the essential adventure of Light.

if this were to happen then an essential degeneration will set in at several levels of granularity. With the emergence of subsequent layers of complexity, systems of feedback are put in place which can either stall such degeneration or strengthen it.

The layer of culture, perhaps a byproduct of any human endeavor, in that human perception, bias, behavior, and action gets stabilized as the norm of that endeavor, hence becomes critical in allowing a human to step into broader vistas of light.

Culture itself, as with any light-based quadrality, can be viewed as a play of Light's properties of knowledge, power, service, and harmony. In designing a human-based organization it is critical, therefore, to allow the manifested sets or the range of elements of knowledge, power, service, and harmony, to increase.

A culture in which Knowledge can express the vastness of the possibility in it, perhaps manifesting as a seeking for deep truths, a mindset of seeing things from multiple points of view, an essential drive toward a synthesis of possibilities, amongst many other such elements, would not only help a human to take the high road in terms of personal choice, but also continue the essential adventure of the expression of fourfoldness.

Similarly, a culture in which Power can continue to express its possibilities, perhaps manifesting as a sense of adventure, of energetically driving toward new possibilities, of leadership, courage, of an innate sense of justice always protecting that or those who are weaker, amongst many other possibilities, would similarly allow personal choice to move toward the direction of Light.

A culture of Service, in which no endeavor is ever thought of as too impossible, where there is a dedication to the task at hand, a deep sense of humility, persistence, attention to detail, of losing of one's personal desire in the work to be done, of having no expectation of reward, amongst other possibilities, will also allow the arc of increasing fourfoldness to continue to advance.

A culture of Harmony, where a recognition that many parts need to come together in work, where there is respect for others based on who they are and not who you want them to be, where a working out of interpersonal obstacles has become systematized, where in fact any obstacle is seen as an opportunity to further ground the potential in Light, will similarly allow personal choice to strengthen the play of fourfoldness.

The interplay between the human-based quadrality and such knowledge-power-service-harmony culture light-based quadrality must become such that there is the natural and easy overcoming of narrowness and organic growth of positive values.

When such layers of emergent light-based complexity reinforce the fourfold possibility within each quadrality then the adventure of Light will continue revealing more and more beauty, love, and possibility.

#31: Toward a Comity of Nations

The choice available to the human-being confers a level of complexity to practical affairs that previous emergences of light-based quadralities did not have to face. This complexity is in no small measure driven by the initial narrowness with which any of us, as humans, is prone to view circumstance. Such narrowness, it can be argued, creates feedback loops that eventually reflect larger scale macro-level dysfunction.

Therefore, it was suggested that culture in any human-based endeavor becomes important in that it allows pathways for personal choice to take the high-road.

Following the arc of fourfold enrichment, it can also be inferred that the sustainability of humankind will be ensured by a balance of development amongst the four sets of properties of Light. This means that countries and global regions must be allowed to be unique and in such a way that their primary emergence contributes to a collective distribution across all four essential sources of innovation or inspiration.

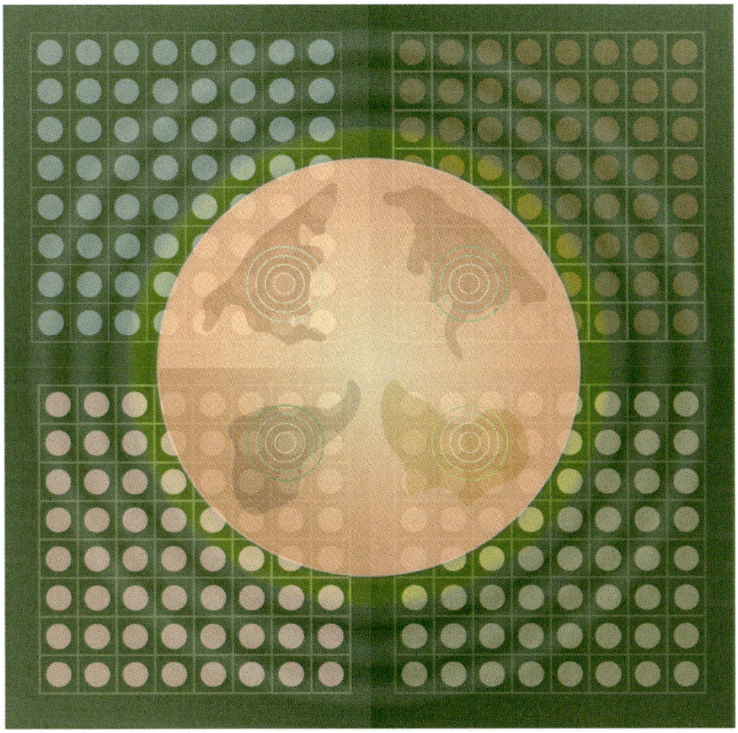

Further, and based on this uniqueness, there must be open and healthy interaction amongst these centers of uniqueness. Practically this means that one single point of view, no matter how it may have benefitted a single country, can never become the sole blueprint for sustainable global development. Countries and regions of the world will have to follow their own genius.

This requires maturity by large organized parts of the world, whether nations or regional blocs. The maturity must be such that the richness implicit in the fourfold properties of Light is allowed to be adequately developed and expressed. In such a scheme it would be normal for different countries to express a unique and weighted combination of some elements from the four sets of properties such that even a hitherto unexpressed element may lead.

Hence it should be expected that countries in their deeper essence are the emergence of some aspect of Light's Knowledge, or Power, or Harmony, or Presence, as the case may be. The possibilities should be vast, and by way of representation only, we will consider here examples of India, Japan, Thailand, and the UK, as suggestive of the four broad and distinct sets of the properties of Light that must be balanced across the globe.

At its best, as perceived historically, India for example may be thought of as having an exceptional capacity for penetrating behind the surface, and further, of meaningfully synthesizing many streams of development. It may be suggested therefore that its primary power perhaps emanates from the family of knowledge.

Japan, at its best as perceived historically, may be thought of as having a strong and noble warrior nature, along with a refined sense of aesthetics, amongst other qualities, hence being representative primarily, of the family of power.

The UK, at its best as perceived historically, may be thought of as having a strong ability to create practical, people-based harmonies, resulting in such things as working parliaments and advanced democracy, thus suggesting that it is representative of the family of harmony.

Thailand, at its best as perceived historically, may be characterized by an exceptional sense of hospitality and service, with attention to detail in the practical arrangement of things, hence suggesting that it is representative of the family of presence.

It is only in following deep levels of authenticity, that that which is unique to a nation can come to fruition. And it is only when the operating principles and ways of being of such genius are recognized as a necessity in the materialization of the infinity resident in Light, that there can be true comity of nations.

#32: The Precision of Cosmic Computational Power

In a Cosmology of Light, it has been proposed that there is a persistent quantum-level computation. The kernel of this cosmic computational power is built on a process of triple quantization involving key mathematical transformations by which the unity in Light at its native state precipitates into vast diversity at light traveling at *c*. The play between 'c' and 'h' — the latter being Planck's constant — is instrumental in the 'physics' of this transformation, and light can be thought to materialize due to the interplay of such physics and the underlying mathematical transformations.

An early output instrumental to the further development of Cosmos is the space-time-energy-gravity light-based quadrality. Through the persistent and underlying quantum-level computation there is a real-time arbitration between what exists in possibility in antecedent layers of Light, and what has expressed itself materially in the physical universe.

Such arbitration has resulted in a stream of light-based quadralities, including the pre-human electro-magnetic-wavearchetype-masspotential spectrum quadrality, the quark-lepton-boson-Higgsboson quantum-particle quadrality, s_Shell-p_Shell-d_Shell-f_Shell atom quadrality, the nucleicacid-polysaccharide-lipid-protein cell quadrality, and post-cellular thought-will-emotion-sensation human-based quadralities, knowledge-power-presence-harmony culture quadrality, and the knowledge-power-presence-harmony united world quadrality.

But these are representative quadralities only. Many intermediate-level pre-cellular quadralities of equal importance, such as belong to the genre of the gluon-W&Zboson-photon-graviton boson quadrality and the cell-based A-C-T-G DNA quadrality, and post-cellular quadralities whose emergence has allowed the creation of any veritable garden of Eden in its richness of plant and animal species has been left out.

While such levels may be left out in a journal of musings such as this, one can be sure that the computational power inherent in the development of the Cosmos implies a level of precision and assures that the chain of light-based quadralities has an infallible logic to it.

#33: Light-Based Singularities

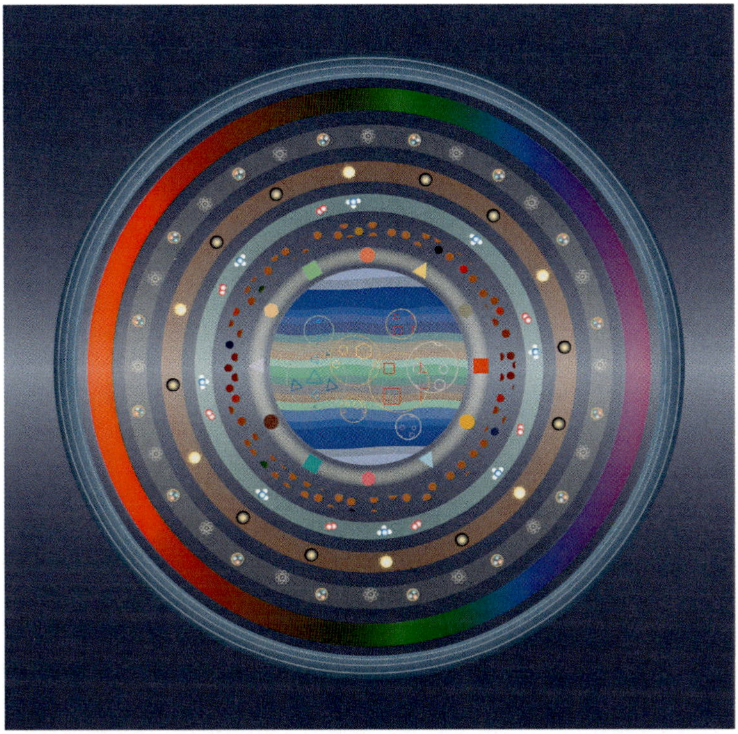

'The Precision of Cosmic Computational Power' made the case that the interaction of the physics and mathematics of Light causes the materialization of Light.

This materialization results in an array of light-based quadralities starting from the space-time-energy-gravity light-based quadrality, through the pre-human electro-magnetic-wavearchetype-masspotential spectrum quadrality, the quark-lepton-boson-Higgsboson quantum-particle quadrality, the gluon-W&Zboson-photon-graviton boson quadrality, the s_Shell-p_Shell-d_Shell-f_Shell atom quadrality, the nucleicacid-polysaccharide-lipid-protein cell quadrality, and post-cellular thought-will-emotion-sensation human-based quadralities, the knowledge-power-presence-harmony culture quadrality, and the knowledge-power-presence-harmony united world quadrality.

Yet if we examine the functioning of each of these it appears that control in these quadralities is largely maintained by the computational process at the Cosmic level. With the emergence of the human-based quadralities, some of this control appears to be transferred to the human. Yet the awareness and range of perception, action, and decision-making are narrow, and largely limited to the container within which a particular individual arises.

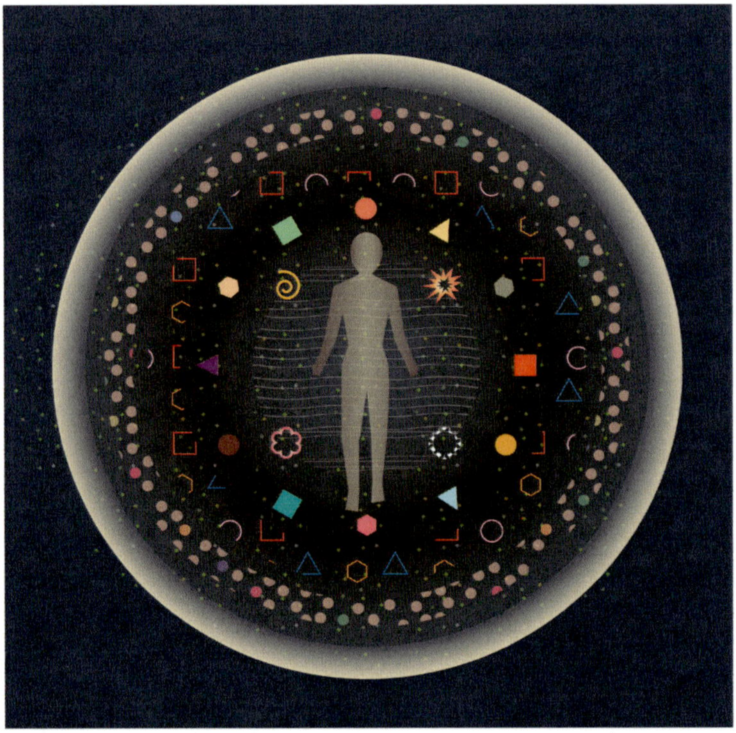

When a human becomes aware of their emergence from the single light-based edifice, when a human can freely enter into the logic of micro and macro space-time-energy-gravity containers, when a human realizes that they are one with the Light, and can freely move, be, and act as Light, then it may be said that the human has been endowed with the computational power that thus far had been the province of the Cosmos only.

Such an endowment, because there is now a full awareness of the nature of Light both within and without, and an awareness that the self is nothing other than Light, would mean that the emerging entity is a Singularity — a Singularity, because, that which has emerged is conscious and One with all of Light.

In a Cosmology of Light, though, by definition, whether there is awareness of the single nature in Light, whatever emerges is nothing other than Light. If such an emergence is a light-based quadrality, so that the fourfold foundation of its existence has fully emerged as in all the cases cited above, then such an emergence may be thought of as a Partial-Singularity.

Most emergences to date, therefore, are partial-singularities.

It is only when the emergence also has full awareness and effective computational power to direct, that it will have become a Singularity.

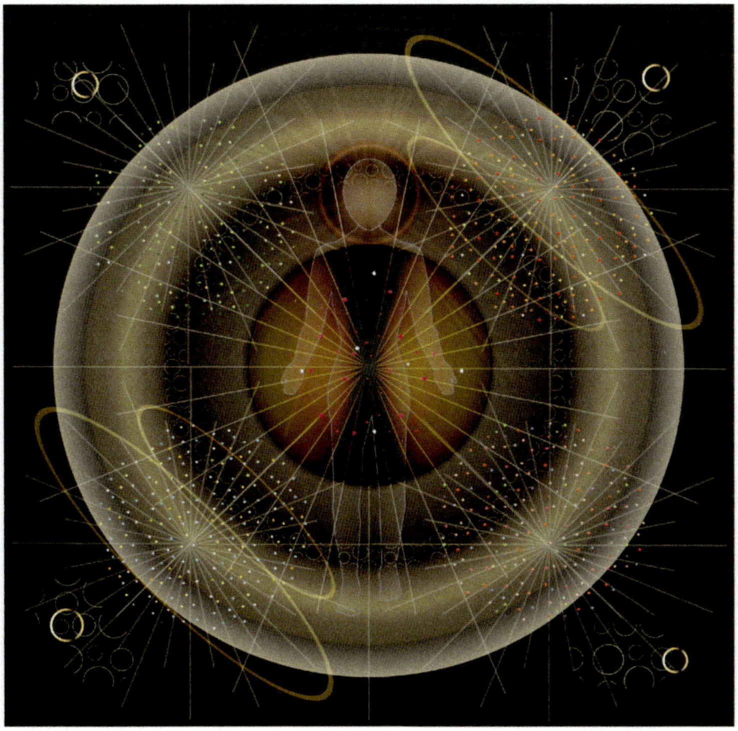

#34: The Significance of Number

Any cosmology should be complete enough that the significance of number is evident as a result of its processes and operations. This is indeed the case in a Cosmology of Light as this musing will illustrate.

Zero (0): The meaning of '0' is encapsulated by the reality envisioned to exist in Light's native state. This is a state of all possibility, in which Light travels infinitely fast. This all possibility can be thought of as an Absoluteness, that then becomes the significance of '0'.

One (1): When Light's native state is perceived slightly differently then the unity and integrality of everything in it become clear. '1' then signifies the concept of Unity.

Two (2): In a Cosmology of Light, it is envisioned that materialization begins when light projects itself at a slower speed. Such a projection signifies a Decision to Create or to Project, and this becomes the significance of '2'.

Three (3): An elaboration of such a projection would reveal three states. There is some original state from which the projection takes place, and then two ways in which the projection would exist — one in which it is static, and another in which it is dynamic. These three states can be thought of as the vertices of a triangle, that becomes a first stability and marks the beginning of the creation. '3' then signifies this Stability or the Beginning of Creation. Further, the 'top' vertex can be viewed as a play of Knowledge. The vertex symbolizing the static aspect can be viewed as a play of Presence. The vertex signifying the dynamic aspect can be viewed as a play of Power.

Four (4): The awareness created within the stable space signified by '3' allows choice and action, and this is how the principle of Harmony begins to be established. So, the addition of the fourth principle within a stability then reflects something of the four properties of Light. Such a material reflection of the four properties implicit in Light signifies a manifestation. '4' hence signifies Manifestation.

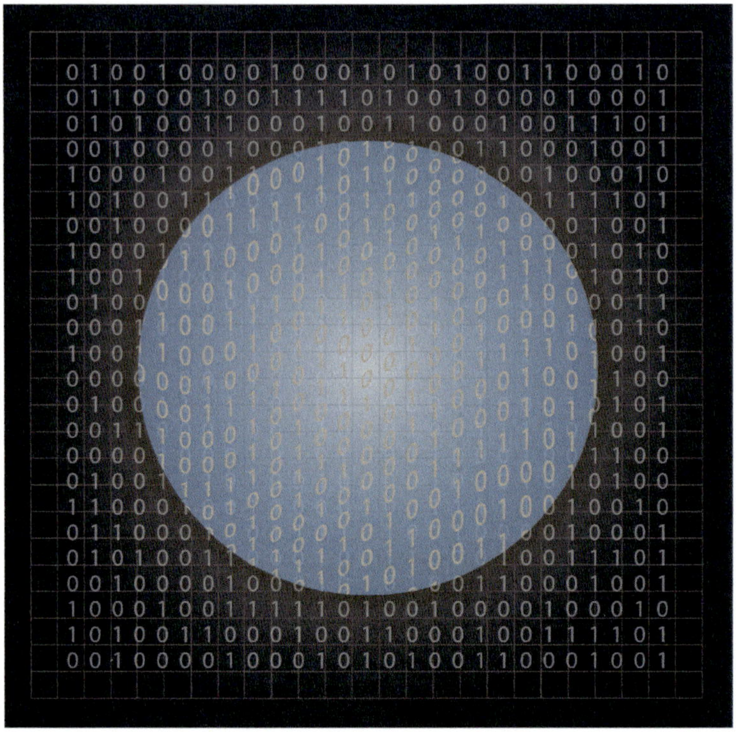

Five (5): '5' can be thought of as the addition of '4' and '1'. The principle of manifestation supported by an underlying unity so that it can repeat itself, again and again, is what power is. '5' then signifies Power.

Six (6): '6' can be thought of as the addition of '4' and '2'. The combination of the principles of manifestation and the decision to create, then signifies creativity. So '6' signifies Creativity.

Seven (7): '7' can be thought of as the addition of '4' and '3'. The combination of stability and manifestation signifies realization. '7' then signifies Realization.

Eight (8): '8' can be thought of as the addition of '4' and '4'. This would be both an inner and an outer manifestation which would signify abundance. '8' then is Abundance.

Nine (9): '9' is the addition of the principles of '4' and '5'. When manifestation and power come together then manifestation continues to boundaries. This dynamic can be thought of as a process of creativity. '9' then signifies Process of Creativity.

Ten (10): '10' suggests that whatever specific potentiality existed in '0' that sought to express itself is now complete. '10' then signifies Completion.

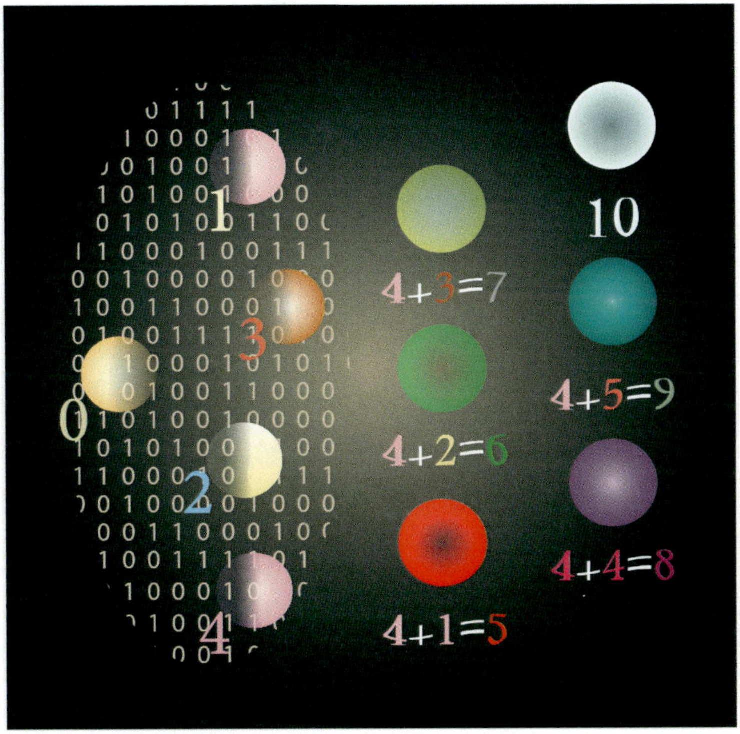

#35: The Fourth Quantization

Light in its native state reveals properties of omnipresence, omnipotence, omniscience, and omniharmony. In a Cosmology of Light, Light can project itself at different speeds and in so doing will reveal a symmetrical face of its fourfold reality.

Were Light to project itself such that it existed at zero speed, then the expression of its infinity would occur in another way.

Its property of Presence would manifest as the absence of light, or as darkness. Its property of Power, unable to influence anything, would become weakness. Its property of Knowledge would be its manifest opposite or ignorance. Its property of Harmony, unable to hold anything in its substance would result in chaos.

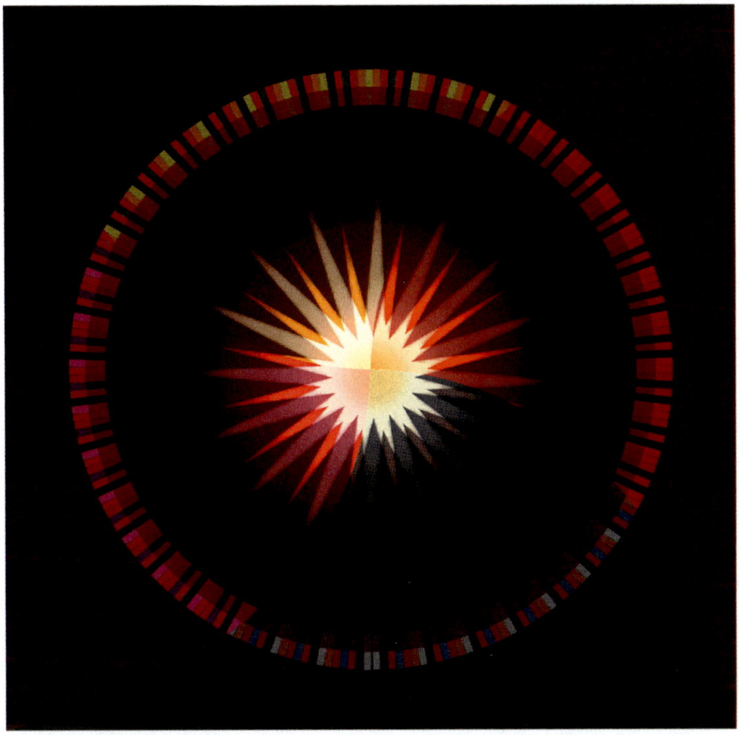

But yet this extreme quadrality of darkness-weakness-ignorance-chaos would have enclosed within it all the infinity of Light. Because of that perhaps, there would be a different kind of influence from this nether state to the layer where Light travels at speed c.

Such an influence would be another quantization, in addition to the envisioned triple quantization — a *fourth quantization*- therefore, and signify a contrary input into the persistent quantum-level arbitration of possibility that architects practical existence.

Hence, not only would there be the input from antecedent layers of Light, but there would be this spur of darkness which by contrary effect perhaps urges emergences to open more to the Light.

#36: The Effect of the Four Quantizations on Space and Time

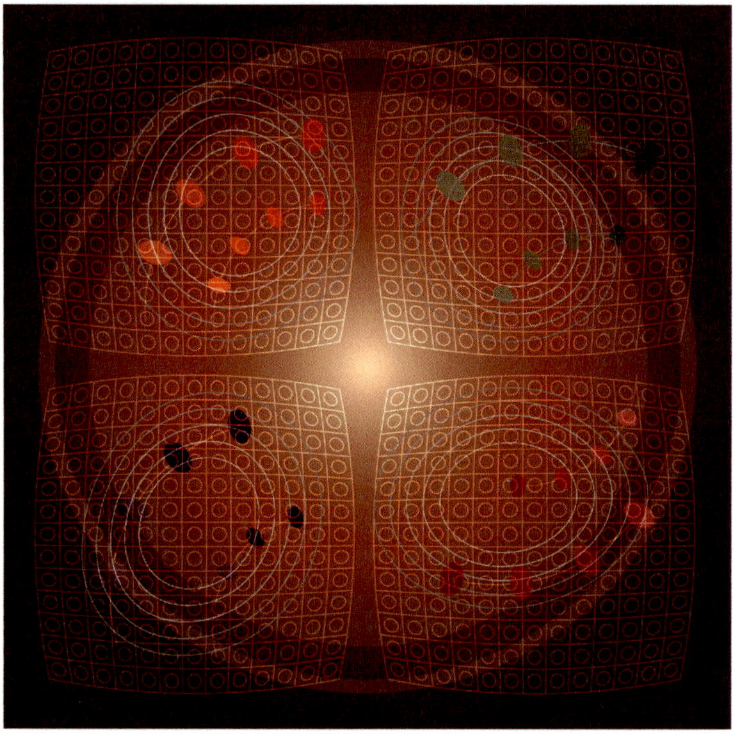

The four quantizations envisioned in a Cosmology of Light can be thought of as four kinds of materializations of light. Each of these materializations is thought to influence the way in which Space and Time are experienced.

'The Logic of Triple Quantization' summarizes three quantizations from Light's native state to the layer in which Light travels at speed c. The first quantization materializes the four properties of Light so that each reveals more of what is hidden in it, and this can be visualized as forming four independent sets. The second quantization further materializes the vast and subtle information in Light by combining elements from the four sets into an infinite number of unique though subtle-seeds. The third quantization materializes the subtle-seeds so that they become the grain of Space and set into motion the play of Time as suggested in the 'Mystery of Space, Time, Energy, and Gravity'. The fourth quantization makes tangible the dark influence from a nether state in which Light is wrapped in darkness.

But each of these quantizations is nothing other than a dynamic of light, and the nature of Light being One, its dynamics can be accessed and made active even in the physical realm regardless of what layer of Light they may be associated with or anchored to.

If it is the first quantization that becomes active, by definition, the purity of Light is at a higher level than compared with the subsequent quantizations, and its effect will be felt tangibly differently than the more materialized Light due to subsequent quantizations.

Hence, the nature of Space and Time will change depending on what kind of quantization is active.

The first quantization relates Light where it travels infinitely fast to some projected layer where it is slower. When Light travels infinitely fast Time will be eternally Now without passage. The infinity of Space will be present everywhere. The influence of such a quantization on the known physical realm will therefore be such that Light will exist with more purity and time will elongate.

With the second quantization the relative 'purity' of Light is diminished, and as suggested in 'Bounds of Existence', the lower bound of the equivalent of Planck's Constant, 'h', will tend away from zero. 'Matter' at that layer will form differently and be more like a field. Time, therefore, required to work out dynamics in a different way, will tend away from Now, but still maintain its relative elongatedness, while Space relatively speaking, will begin to collapse. The influence of such a quantization will therefore set up a different kind of Light, Time, and Space on the physical realm.

With the third quantization, the trend will continue and with the creation of subtle-seeds, Light is no longer experienced as One, and Time and Space get further fragmented. The influence on the physical realm will be felt as relatively less pure light, though still purer than when Light travels at speed c.

With the fourth quantization, there is a strong influence of the darkness-weakness-ignorance-chaos quadrality.

The influence of these four quantizations can therefore set up substantially different macro-cycles in which Light, Space, and Time are different. There may be some correlation with the esoteric cycles embodied by the Satya-Yuga, Treta-Yuga, Dvapara-Yuga, and Kali-yuga which each have a diminished play of Light, and that apparently govern a cyclic process of Time.

#37: Determinism to the Cycles of Time?

'The Effect of the Four Quantizations on Space and Time' suggested that the effective or invoked quantization during a period or age will have an impact on how time itself is experienced. The four types of quantization were further suggested to have some relationship with the esoteric ages known as Satya or Golden Age, Treta or Silver Age, Dvapara or Bronze Age, and Kali or Iron Age, respectively.

But further, in some circles, there are calculations on how long each of these ages actually is. In this musing, one such calculation will be reviewed from the point of view of a Cosmology of Light.

The Significance of Number suggested that in a Cosmology of Light "number" is encapsulating a concept. But beyond that, it can also be observed that there is a sequential logic to the numbers such that a lower number has to occur for the concept embodied by the higher number to make sense.

Hence, it can also be observed that the number 7 signifies a 'realization' and that the sequence 1 through 7 is a logical sequence culminating in 'realization'. It can also be observed that the sequence 7 through 10 culminates in 10, signifying 'completion'. 7 has to be reached before 10 can be, and further, 7 is the start of a new journey that can result in 10. 7, therefore is also a bridge-number.

Numerically there is also an interesting pattern that is observed when taking these two sequences independently.

Hence 1x2x3x4x5x6x7 = 7x8x9x10 = 5040.

But '5040' is known to be the main part of the fundamental block, with the dusk and dawn of that block being a tenth of the main part respectively. Hence, 504 (dawn)+ 5040 (main part of a block) + 504 (dusk)= 6048, which in ancient Indian texts is considered to be the number of years in the Kali-Yuga.

But the Kali-Yuga is when time is harried and experienced in a rushed way so that the sequence 1 through 7 is of an entirely different nature than the sequence 1 through 7, in any of the other ages. As per the common calculation, 1 through 7 is perhaps experienced 4 times slower in the Satya-Yuga, three times slower in the Treta-Yuga, and two times slower in the Dvapara-Yuga. Or alternatively, in the Satya-Yuga there are two separate 1–7 and 7–10 cycles. In the Kali-Yuga there may be only one 1–7 cycle.

Such calculations though suggest a mechanistic determination of time and cycles, and perhaps may exist if Light were unchangeable in practical affairs. The whole point of a Cosmology of Light though is to suggest that such determinism can in fact be altered by opening more to Light, here and now.

#38: A Garden of Eden in the Fourfold Scaffolding

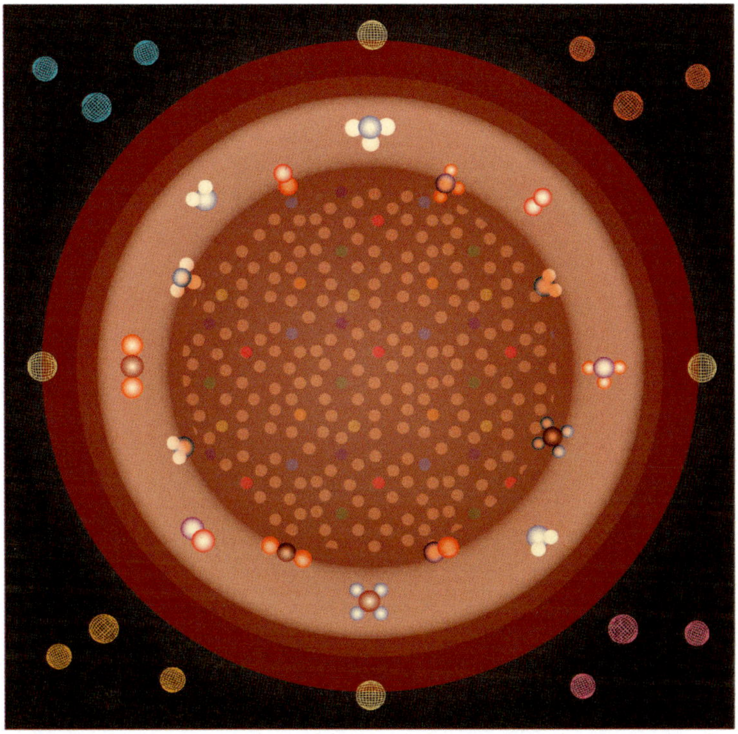

'The Precision of Cosmic Computational Power' made the case that there are a long series of light-based quadralities, one building on another, that have contributed to the tremendous diversity of material existence.

The cellular-based molecular plan quadrality precipitated a harmony of possibility from the infinity in Light, which can be thought of as the veritable materialized seed from which vast diversity in living form came forth.

But this all started from a prolific expression of plant life. Plant life and simple pre-human life, not yet subject to the vicissitudes of choice, in turn, ensure rich ecosystem-services development.

Hence, food, raw materials, genetic resources, biogenic materials, medicinal resources, energy, climate regulation and carbon sequestration, waste decomposition and regulation, water and air purification, disease control, amongst many other such services, all emerge naturally with the emergence and interaction of matter, plant, and simple animal life.

Such ecosystem services development ensures that the scene has been set and that a Garden of Eden comes into being, all dictated by the fourfold possibility existing in Light. Perhaps even a Garden of Eden "being" — an interim and full singularity in the endless expression of fourfoldness — required the emergence of all previous matter, plant, and animal life to express itself.

The question now, with the emergence of the human, and human-based organization, is whether the integrity of the fourfold scaffolding, billions of years in the making, will be allowed to remain in place. This perhaps will depend on the nature of quantization that effectively comes into play.

#39: The Human-Based Complex Adaptive System

In the 'Basis of Complex Adaptive Systems', the case was made that it is the layering of light-based quadralities that in fact creates the phenomenon of complexity. This is due to the fact that agents in any particular layer can now not only interact and influence agents in their own layer but in adjacent layers as well. The resultant and unexpected feedback loops are the cornerstone of complexity.

When considering human-based complex adaptive systems (CAS), there are several layers that create the complexity and the dynamics that CAS is known for.

There is of course the layer due to the human light-based quadrality. But then there are also a couple of foundational layers due to the fourfold living-cell that is the basis of all life, and the containing natural environment with its plants, animals, and the plethora of built-up ecosystem services as referred to in 'A Garden of Eden in the Fourfold Scaffolding', Additionally, there are also the layers that come into being due to efforts to create culture, — whether in corporate, social, economic, or political realms — and to further organize civilization at the national and international levels as referred to in 'Toward a Comity of Nations'.

Human-based CAS is therefore practically a result of the interaction of at least five light-based quadrality layers.

While light-based quadralities, as an emergence from the infinite potential resident in Light's native state, seek naturally to materialize a greater fourfold richness, this trend is easily compromised in human-based CAS due to the vicissitudes of human choice.

Such choice, often based on a narrow and short-term conception of self, causes unforeseen feedback loops between layers, creating a reversal of the growth of fourfoldness in and across layers. Formidable global problems such as systemic pollution, Climate Change, COVID-19, financial recessions, lack of inclusive growth, unresolved human illness, resource depletion, biodiversity loss, urban sprawl, to mention a few, arise in its wake.

Instead, then, of seeing the real problem of the depletion of fourfoldness, we often put in place short-sighted fixes local to a layer, which in turn spawns a whole new set of unforeseen feedback loops. What is needed is a clear movement away from the limitations of the fourth quantization, a clear understanding of the power of Light and the effect of the appropriate quantization, and an invoking of higher quantization or more holistic choice at the human-level.

#40: Necessary Conditions for World Peace

The creation of World Peace is clearly a formidable task and if we were to frame the task in terms of complex adaptive systems (CAS) theory, we could take the quintuple-layered 'Human-Based Complex Adaptive System' as a starting point.

In such a framing the primary layer is due to the human light-based quadrality. This is so because the vicissitudes of human choice can easily set up a number of destructive feedback loops with adjacent layers. The default action can always originate from a fourth quantization impetus, where rather than opening to higher layers of Light, the thought-will-feeling-sensation human light-based quadrality consistently reinforces short-sightedness and destructive dynamics that chip away at the fourfold integrity of the other layers.

Hence, the foundational fourfold living-cell layer that has naturally emerged through billions of years of evolution and is programmed with sophisticated defense mechanisms to counter a vast range of negative circumstances, can get compromised due to lifestyle choices, become progressively acidic, and trigger irregular cell behavior or even cell death.

But this is by no means inevitable. If other quantization is tapped into then a constructive feedback loop between the layer of foundational living cells and the human-layer can be put into place, and rather than weakness being the outcome of human action, strength, flexibility, resilience can become the natural outcome. If humans were to go this route, then a different kind of natural adaptation at the cellular level would mean that negative environmental influences, whether due to electromagnetic or chemical pollution, or due to unexpected strains of viruses, amongst other perceived negative possibilities, could lose their debilitating power, and this would set in place a necessary condition to enable personal peace, that is itself a necessary condition for World Peace.

If the right quantization were in place, then different feedback loops would exist between natural environments and humans, and instead of destroying the very ecosystem services our lives depend on — ranging from water and air purification; food and raw materials generation; construction of genetic resources, biogenic materials, and medicinal resources; creation of energy, climate regulation, and carbon sequestration; waste decomposition and regulation; disease control; amongst many other such services — these would thrive and also create a necessary condition for World Peace.

Such ecosystem services also create rich work possibilities and can become a more deliberate and important part of industry in all parts of the world, enabling global economic life in a holistic way driven by the wiser rhythms of Nature. Instead, we have lost balance, and now often strip down the wealth of forests, and in a sleight of accounting think we have grown richer by equating the weight of the dead wood to money. This brings us to the often-destructive feedback loops involving culture, whether at the corporate, social, economic, or political areas, also due to the invoking of the dark quantization when as human beings we continue to act from smallness.

If instead, we allowed the richness of the fourfold properties of Light to continue to manifest, then balanced environments driven by knowledge power, service, and harmony, would create richer cultural realities.

Knowledge perhaps would manifest as a seeking for deeper truths, a mindset of seeing things from multiple points of view, an essential drive toward a synthesis of possibilities. Power perhaps would manifest as a sense of adventure, of energetically driving toward new possibilities, of leadership, courage, of an innate sense of justice always protecting that or those who are weaker. Service would awaken a sense in which no endeavor is ever thought of as impossible, where there is a dedication to the task at hand, a deep sense of humility, persistence, attention to detail, of losing of one's personal desire in the work to be done, of having no expectation of reward, amongst other possibilities. Harmony would engender recognition that many parts need to come together in work, deeper respect for others based on who they are and not who you want them to be, and a sense that obstacles are opportunities to further ground the potential in Light, amongst other possibilities.

The reinforcing loop between this layer of culture and the human would create environments where higher quantization is naturally invoked, thereby also creating a necessary condition for World Peace.

The fifth layer has to do with the development of civilization so that countries and global regions are allowed to be unique and in such a way that their primary emergence contributes to a collective distribution across all four essential sources of innovation deriving from Light's fourfold properties.

Today's set of feedback loops emanating from a fossilized view of possibility at the human light-based quadrality level instead sets in place the contrary and forces regions of the world to "develop" along certain lines only. Invoking higher quantization would mean that based on emergent uniqueness, there must be open and healthy interaction amongst national or regional centers of uniqueness. Practically this means that one single point of view, no matter how it may have benefitted a single country, can never become the sole blueprint for sustainable global development. Countries and regions of the world will have to follow their own genius. This would put in place another necessary condition for World Peace.

The dynamics of complex adaptive systems are such that even small parts can have a profound influence on systemic reality as a whole. Individuals, consistently invoking higher quantization to even open to the native reality of Light, can become the drivers of extraordinary change and usher in the reality of the prerequisite conditions.

These conditions, emanating from different layers of light-based quadralities, acting in concert and constructively reinforcing each other, would move us closer to a reality of World Peace

Let us hope that in the years to come this is indeed what is done.

#41: The 'Imaginary' at the Heart of Mathematical Reality

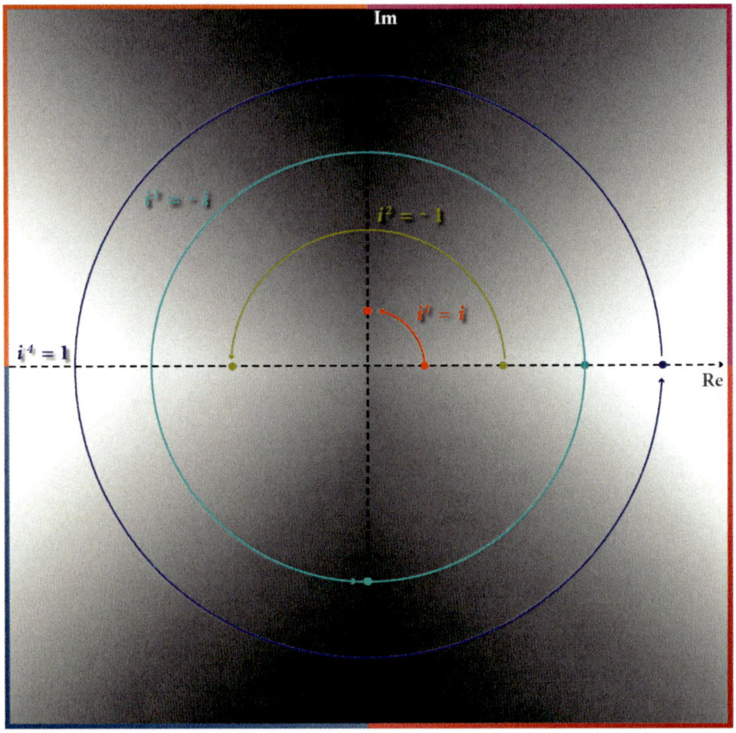

The imaginary number, *'i'* (square root of -1), is an apparent paradox at the heart of mathematical reality. The square root of -1 is, after all, given accepted rules of mathematics an impossibility. Yet without 'i' many equations would never be solved, and many practical technologies would never have come into being.

And so, it is not only a convenient precedent when considering a Cosmology of Light, that consists of different layers of light antecedent to our known layer where light travels at speed 'c', but further, the existence of 'i' in any mathematical rendering gives place to something that clearly influences practical reality, though may not necessarily be considered as part of practical reality.

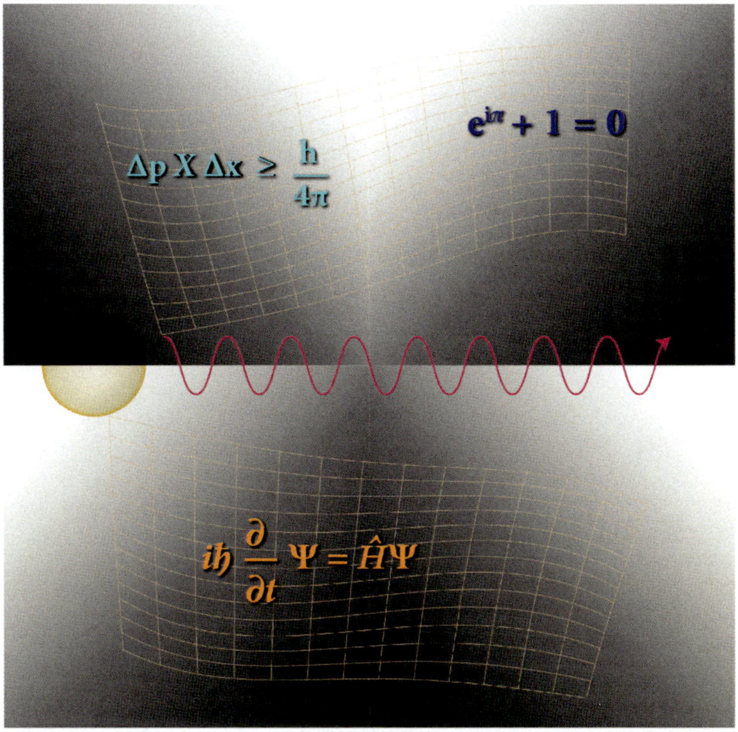

In the next few musings, the relationships between Schrodinger's Equation, Heisenberg's Uncertainty Principle, Euler's Identity, and a Cosmology of Light will be suggested. In this rendering, just as there is a significance to every number in a Cosmology of Light, 'i' in fact, will also be seen to be a necessary device to account for planes of reality created by Light traveling at speeds other than 'c'.

Such a rendering will allow different, Cosmology of Light-based interpretations to the body of equations that inform quantum physics and computation, amongst other foundational areas.

#42: Insight Into Schrodinger's Wave Equation

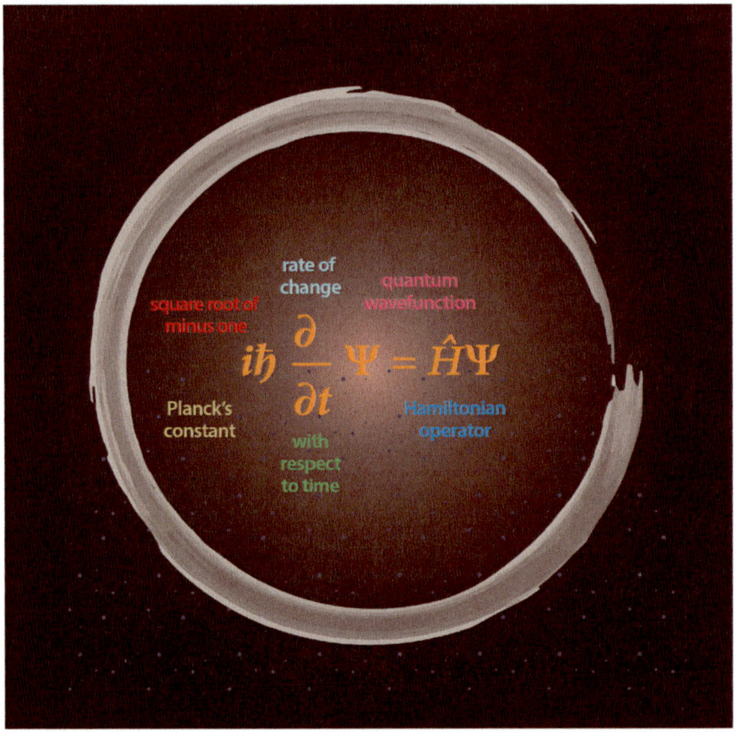

In a Cosmology of Light, any kind of materialization is the result of a persistent quantum-level computation. This persistent computation can be thought of as an arbitration involving information from multiple layers of Light.

Such arbitration as suggested in the musings here creates *Matter* or materializes the information that exists in Light's native state, in the layer where light travels at speed 'c'. The process of such materialization therefore must provide insight into Schrodinger's Wave Equation, that after all is nothing other than a depiction of matter as a wave.

Considering the left-hand side of Schrodinger's Wave Equation:

1. As discussed in 'The 'Imaginary' at the Heart of Mathematical Reality', 'i', the square root of -1, is considered to be imaginary and connects an imaginary plane with the real plane. In other words, 'i' can be thought of as connecting the antecedent layers of Light, to the layer of light where it does travel at speed 'c'.
2. In 'Bounds of Existence', the case is made that just as light traveling at speed 'c' is an intentionality and sets an upper bound to material existence, so too 'h' or Planck's Constant, inversely proportional to 'c', sets a lower bound and mathematically suggests how quanta will materialize. So, if light were to travel faster than c, thereby increasing the upper bound in that layer of existence, then the lower bound 'h', will have to correspondingly become smaller. But this is precisely what we see in Schrodinger's Wave Equation above, where 'h' is divided by '2 x pie', depicted by $\hbar$, to make it smaller, implying too that the 'wave' in the equation also exists in a layer other than the one defined by light traveling at 'c' only.

3. Finally, the rate of change of the wavefunction, is only another way of suggesting that there is a persistent quantum-level computation or arbitration involving all the layers of Light.

Matter, hence, can be thought of as the product of all three of the actions captured by the left-hand side of Schrodinger's Wave Equation.

Further, considering the two terms on the right-hand side of Schrodinger's Wave Function:

1. The Hamiltonian Operator captures the active energies involved in the arbitration.
2. The Wavefunction, subject to the energies summarized by the Hamiltonian Operator, will create the next iteration of Matter.

While the terms on the left-hand side can easily be seen to be emergent from a Cosmology of Light model, the terms of the right-hand side appear to be more general, summarizing energy states, rather than underlying ontology.

Richard Feynman is known to have said that there is no way that Schrodinger could have derived such an equation — it must have come straight out of his head. On the contrary, this series of musings suggest a framework from which an equation like Schrodinger's Wave Equation could indeed have emerged.

#43: Heisenberg's Uncertainty Principle

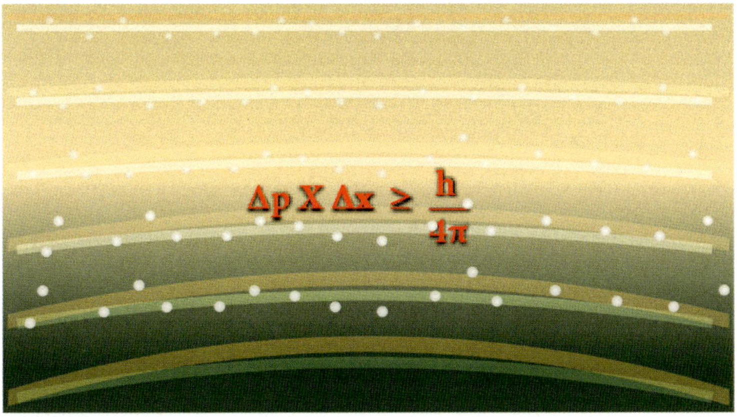

If a group of molecules is cooled to absolute zero so that no more thermal energy exists, the molecules will still continue to vibrate. This vibration is attributed to Heisenberg's uncertainty principle. The equation above depicts such uncertainty in knowing any pair of attributes such as momentum (p) and position (x) at the atomic level.

In a Cosmology of Light though, there is the arbitration of possibility involving information at multiple layers of light. Information from layers of light that exist at speeds greater than 'c' precipitate in relatively slower-moving layers of light through the device of quanta. Such meta-layer information can be thought of as a presiding function that will fulfill itself by using any means at its disposal in the slower moving layer of light.

Such leeway in 'using any means at its disposal' appears as 'uncertainty' in the slower moving layer of light.

But further, as suggested in the piece on 'Schrodinger's Wave Equation', 'reducing' the value of Planck's Constant (h), already implies that another layer of light is operative. And so, when interpreted in this manner, then Schrodinger's Wave Equation is drawing attention to the arbitration involving multiple layers of light, while Heisenberg's Uncertainty Principle is drawing attention to the fact that the process of materialization of the meta-level function can take place in different ways.

#44: Euler's Beautiful Equation

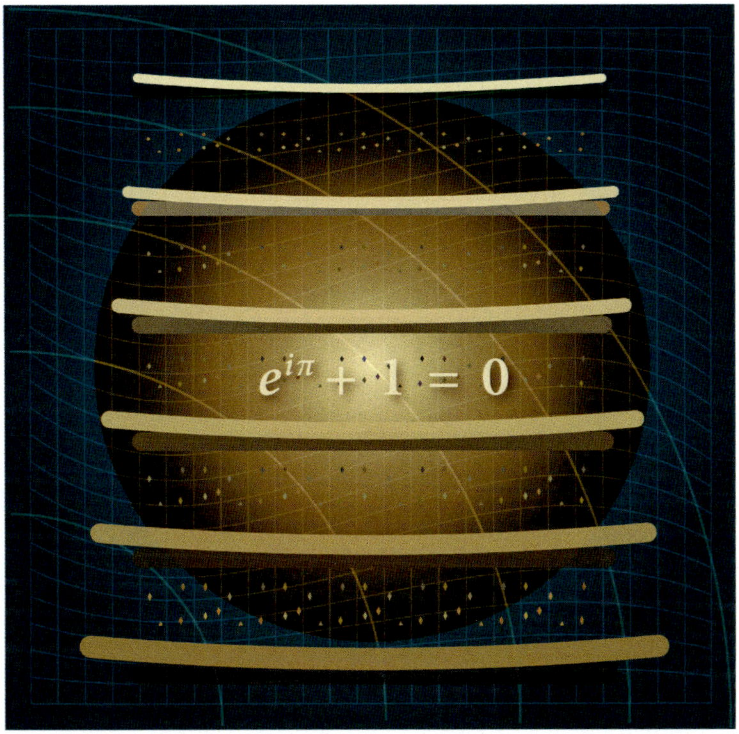

Euler's Identity is considered to be amongst the most beautiful of equations in the world because it connects together 5 fundamental concepts or numbers in mathematics — 0, 1, 'i', 'e', 'pi' — in a simple equation.

But what could Euler's Identity signify in a Cosmology of Light? Just as there are meaningful interpretations for the imaginary number, 'i', Schrodinger's Wave Equation, and Heisenberg's Uncertainty Principle, so too there is a meaningful interpretation for Euler's Identity, as will be explored in this musing.

'The Significance of Number' suggested the deeper meaning of each number from a Cosmology of Light point of view. '0' is considered to be the starting point, within which all possibility exists. '1' is unity and is another way to view the absoluteness and all possibility of that which is in '0'. '1' hence is telling us something additional about the fundamental nature of existence.

'2' is considered to be the decision to create, as made evident from the projection of Light from its native state to some other state. '3' signifies the first stable form and signifies the beginning of creation. It is in considering the process between '2' and '3' which gives insight into 'e', whose value is pegged approximately at 2.71828.

'e' is associated with the exponential function, and when viewed between '2' and '3' can be understood as that in which the essential polarity created by '2' becomes a veritable device for experimentation or adventure, a crucible of fire of sorts, which eventually brings forth or yields to that which was being created within it, into a form or stability associated with '3'.

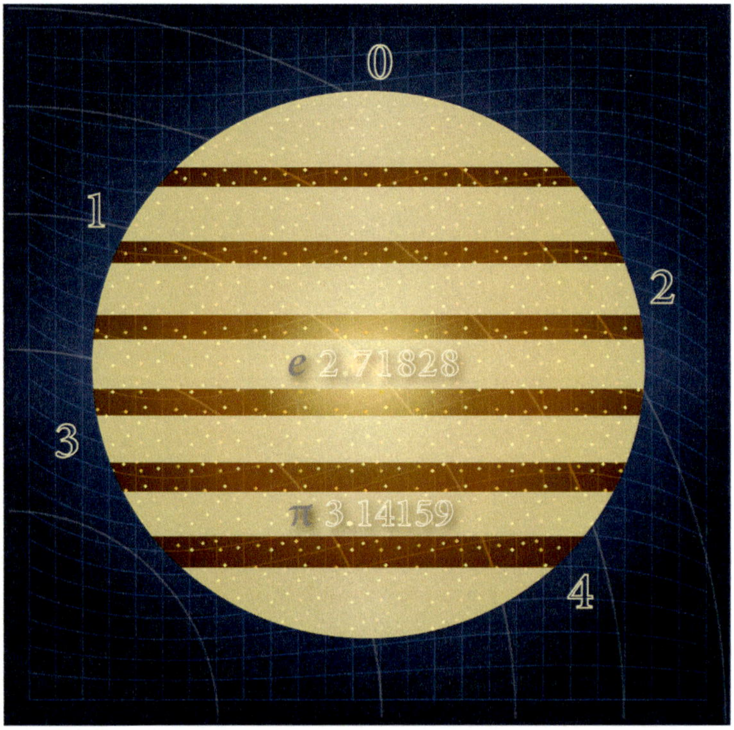

Similarly, insight into 'pi' can be gained by realizing that its value, approximately 3.14159, lies between '3' and '4'. '4', displaying explicitly the four properties innate within Light, suggests manifestation. Between creation (3) and manifestation (4) therefore, 'pi' signifies a constructive process in which stability, in a manner of speaking, is subjected to dynamism, to bring forth a fuller manifestation.

But also, when considering the 5 mathematical concepts in Euler's Identity, the '0', '1', and 'e' — which are meta-physical giving insight into process in antecedent layers of light — are connected to 'pi' that suggests materialization in the layer where light travels at speed 'c', and this connection as considered in the musing on imaginary numbers takes place through the imaginary number, 'i'.

Hence, in a Cosmology of Light, Euler's Identity is summarizing a process of materialization of information existing in antecedent layers of Light, in the layer where light travels at speed 'c'. This process takes place by '0' yielding to '1' yielding to '2' yielding to '3' yielding to '4' by appropriately leveraging 'i', 'e', and 'pi', respectively.

#45: Euler's Lullaby — Cosmic Symphony Paradox

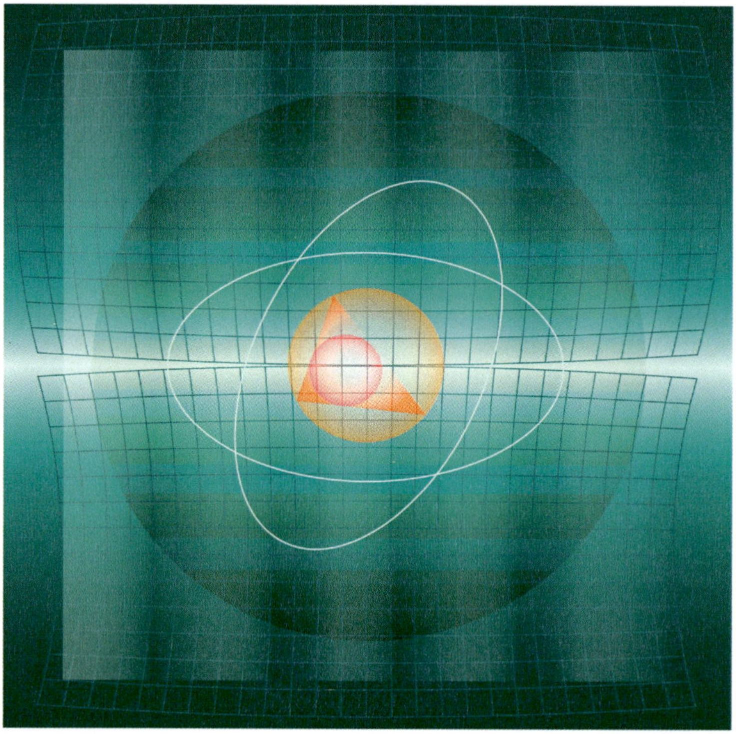

'Euler's Beautiful Equation' suggested an interpretation linking 0, 1, 'i', 'e', and 'pi' in a Cosmology of Light, and Euler's formula (related to Euler's Identity) charts out 'e', 'i', and 'pi' (or an angle depicted as multiples of pi), as an endless wave.

The attraction of such an endless wave is that no matter how much the angle increases (depicted by the horizontal axes above), the value as per Euler's formula will always fall within a set upper and lower bound (depicted by the vertical axes).

For many quantum physicists, this is perhaps utopia, because if the bounds of the axes are set between 0 and 1, then as per the accepted interpretation of quantum mechanics, the value always falling between 0 and 1, becomes a probability.

Voila! Euler's formula now becomes the means by which any richness in terms of entanglement or superposition or any other quantum phenomenon that happens behind the veil, can be equated to some probabilistic outcome happily measured on the surface. Hence Euler's Formula becomes a means to depict everything that goes on behind the quantum veil.

But in a Cosmology of Light, if the undulating wave is used as a probabilistic mechanism only, it becomes nothing other than a *lullaby* that veritably coaxes the aspiring explorer into a state of sleep.

Instead, as suggested by 'Euler's Beautiful Equation', the undulating wave is not a probability function, but rather, is constantly connecting possibility in antecedent layers of Light with the layer where Light travels at speed c. Nothing that happens in the material layer (where light travels at c) can be determined without the arbitration between layers suggested by the action of the undulating wave.

This constant arbitration, following a fourfold enrichment, reveals therefore an ever more full *cosmic symphony*. Seeing it as so changes the way quantum phenomenon is interpreted and has profound implications for quantum computation, and a host of technologies that are in turn imagined to exist on that.

46: Generation of Sound & Music

'Euler's Lullaby-Cosmic Symphony Paradox' used the metaphor of music in its effects on audiences. This musing will focus on some aspects of sound primarily from the perspective of the 'Significance of Number' in a Cosmology of Light.

Light in its native state is imagined to exist infinitely fast, also displaying properties of omniscience, omnipotence, omnipresence, and omniharmony or omninurturance. This omniscient-omnipotent-omnipresent-omniharmonious quadrality can also be imagined to have a primal sound that summarizes the infinite potentiality resident in Light, and perhaps that is the most mysterious of sounds referred to as OM.

But further, as summarized in 'The Significance of Number', the essential unity (1) projecting itself as a creative configuration of unity (2), that is 1 becoming 2, may also be thought of as an 'octave'. Everything in '1', is through the logic of light-based quadralities, materialized as a complete though different face of itself in '2'.

'2' becoming '3' so that the decision to create signified by '2' results in a first stability summarized by '3', is in music theory a 'fifth'. And '3' becoming '4', is the movement from some stability into a first complete manifestation, but can also, in music theory be referred to as a 'fourth'.

Hence, we can begin to see how the meaning of projections of Light, encapsulated by the significance of number, can in the relationship of ratios such as 1:2 (octave), 2:3 (fifth), 3:4 (fourth), also be the basis of music.

#47: An Original Photon

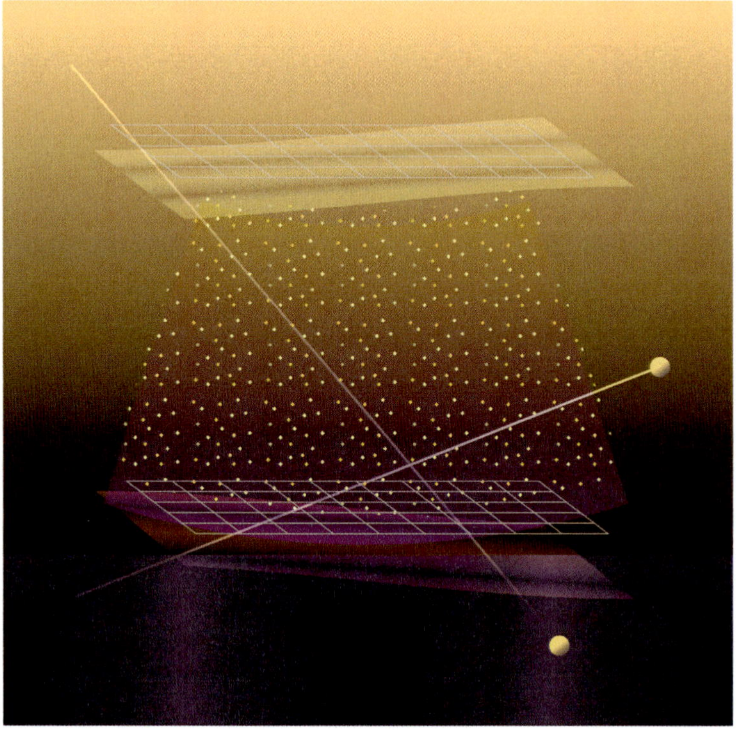

'The Fourth Quantization' proposed that if Light were to project itself at zero speed, it would in effect swallow up all potentiality to materialize a Darkness-Weakness-Ignorance-Chaos quadrality. This would be a persistent influence in the process of quantum-level arbitration that determines reality where Light travels at speed c, aka our known physical universe.

But in another way of looking at it, any quanta are a bridge mechanism connecting one layer of light with another. The common name for a quantum of light is a *photon*. But if in some act Light in its native state projects itself at zero speed thereby creating a first Darkness-Weakness-Ignorance-Chaos quadrality, that too is nothing other than a quantum of light or a photon materializing the opposite of Presence-Power-Knowledge-Harmony.

It, therefore, can be thought of as a contextual or an Original Photon, perhaps one amongst a few such photons, instrumental in how the details of further materialization will take place.

#48: Other Cosmic-Sized Photons

'An Original Photon' suggested that Light projected at zero-speed creates a layer of reality animated by a Darkness-Weakness-Ignorance-Chaos quadrality. The information held together as it were in a macro-sized packet to create a layer replete with such fourfold properties, is therefore nothing other than a cosmic-sized photon.

But there are other cosmic-sized photons that similarly hold information in a pregnant embrace until such point as it can be released to create the fourfold reality of a unique layer of light.

Hence, light projected at speed c can be thought of as such a cosmic-sized photon containing within it the information to create a physical universe. But as described in 'Materialization of Light', so too, there would be additional cosmic-sized photons that created the field-like and wave-like realities that preceded the particle-like reality of the known physical universe.

#49: A Directory of Photons

In 'Other Cosmic-Sized Photons', it was suggested that light projected at different speeds would create cosmic-sized photons. Hence light projected at zero speed creates a photon responsible for the Darkness-Weakness-Ignorance-Chaos quadrality, just as light projected at c, and at speeds between light's native state where it exists infinitely fast, and c, create other cosmic-sized photons.

But when light is projected at c it precipitates the big bang and the appearance of a composite space-time-energy-gravity quadrality. This space-time-energy-gravity quadrality is nothing other than a packet of light, of the subtlety of information embedded in antecedent light expressing itself in a cohesive and significant frame, and therefore, *can also be thought of as a photon.*

But then, by this logic, all subsequent light-based quadralities are also nothing other than a composite packet of light, *a photon,* expressing information embedded in antecedent light in a more structured material form.

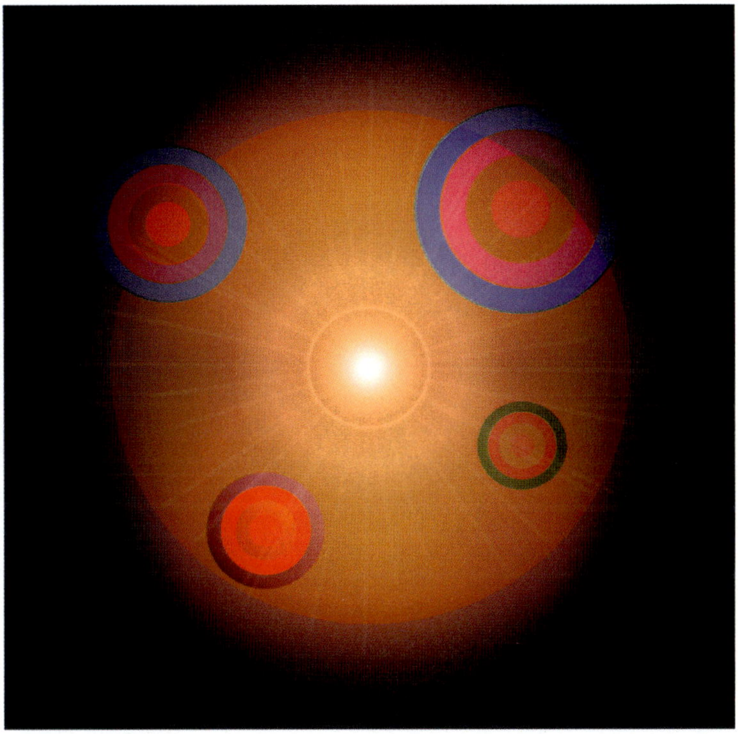

Hence, light-based quadralities starting from the space-time-energy-gravity light-based quadrality, through the pre-human electro-magnetic-wavearchetype-masspotential spectrum quadrality, the quark-lepton-boson-Higgsboson quantum-particle quadrality, the gluon-W&Zboson-photon-graviton boson quadrality, the s_Shell-p_Shell-d_Shell-f_Shell atom quadrality, the nucleicacid-polysaccharide-lipid-protein cell quadrality, and post-cellular thought-will-emotion-sensation human-based quadralities, the knowledge-power-presence-harmony culture quadrality, and the knowledge-power-presence-harmony united world quadrality, are also in a manner of speaking *photons*.

#50: Stillness of Light

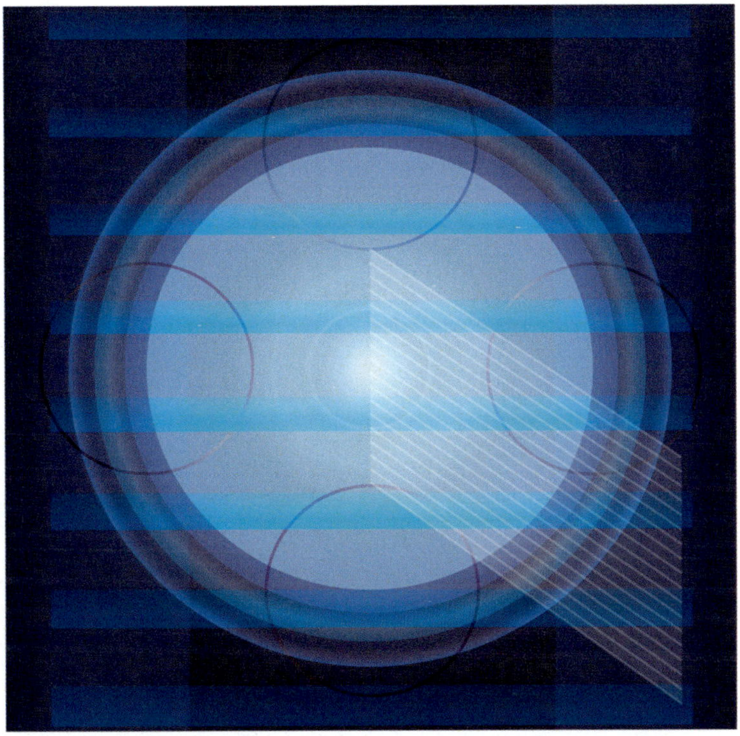

In a Cosmology of Light, Light in its native state has been suggested to travel infinitely fast. But this infinity in speed is relative to the known speed of c, in which the physical universe has arisen.

Considering light to travel infinitely fast allows us to imagine a native reality of omniscience, omnipotence, omnipresence, and omninurturance.

In another way of seeing though we can imagine that Light in its native state is *Still* and that all other projected speeds are just a way for the dynamics of quanta, and therefore of time, space, energy, and gravity to come into being so as to create material realities.

Even at c, there appears to be stillness: and this of a kind that allows another type of dynamism to come to the fore.

Such stillness perhaps then, lets us glimpse something of the eternal mystery that is Light.

Relevant Background and Follow-up Information

The Author's Early Books

1. The Flowering of Management
2. India's Contribution to Management

The Fractal Series

1. Connecting Inner Power with Global Change: The Fractal Ladder
2. Redesigning the Stock Market: A Fractal Approach
3. The Flower Chronicles: A Radical Approach to Systems and Organizational Development
4. The Fractal Organization: Creating Enterprises of Tomorrow

The Cosmology of Light Series

1. A Story of Light: A Simple Exploration of the Creation and Dynamics of this Universe and Others
2. Oceans of Innovation: The Mathematical Heart of Complex Systems
3. Emergence: A Mathematical Journey from the Big Bang to Sustainable Global Civilization
4. Quantum Certainty: A Mathematics of Natural and Sustainable Human History
5. Super-Matter: Functional Richness in an Expanding Universe
6. Cosmology of Light: A Mathematical Integration of Matter, Life, History & Civilization, Universe, and Self

The Application of Cosmology of Light Series

1. The Emperor's Quantum Computer: An Alternative Light-Centered Interpretation of Quanta, Superposition, Entanglement and the Computing that Arises from it
2. The Origins and Possibilities of Genetics: A Mathematical Exploration in a Cosmology of Light
3. The Second Singularity: A Mathematical Exploration of AI-Based and Other Singularities in a Cosmology of Light
4. Triumph of Love: A Mathematical Exploration of Being, Becoming, Life, and Transhumanism in a Cosmology of Light

The Artistic Interpretation of Cosmology of Light Series

1. The Mandala Illustrated Story of Light
2. Musings on Light: A Meditative, Non-Mathematical Summary of a Cosmology of Light

Note on Genesis of Books

In the earlier stage I wrote 'The Flowering of Management' and 'India's Contribution to Management'. The impetus for both these books was similar in that they were reactions to the environment that I was placed in at the time. When I first began working in the corporate world the reality of the environment struck me as decidedly anachronistic. I had a different sense of what life could offer and wrote 'The Flowering of Management' to capture aspects of a vision I thought corporations and money existed for. Similarly, when I wrote 'India's Contribution to Management' it was the result of the dissatisfaction I experienced when confronted with the prevalent interpretation of India. This was precipitated by my working with a US-based company, A.T. Kearney, in India. I sought to reverse that interpretation with 'India's Contribution to Management' which aimed to capture my understanding of the essence and deeper capacity of synthesis of India.

The next phase was marked by a strong interest in fractals that primarily stemmed from my beginning to see similar patterns in seemingly distinct areas of life. I wrote 'Connecting Inner Power with Global Change: The Fractal Ladder' as a theoretical framework of fractals. The fractals that I envisioned included the added complexities of emotional and thought components. This was followed by 'Redesigning the Stock Market: A Fractal Approach' which was an application of the theoretical fractal framework to the then recent global financial crises of 2008. 'The Flower Chronicles' sought to make the gist of the ubiquitous fractal I had described in the previous two books easily graspable at the visceral level primarily through many practical examples spanning diverse walks of life. 'The Fractal Organization: Creating Enterprises of Tomorrow' was a comprehensive summary of the fractal framework that included the basic theory, the applications, and a practical field guide that had been developed while I was working at the Organizational Development group at Stanford University Medical Center.

The most recent phase has focused on creating a mathematical framework to take the previously developed fractal framework further. The development of such a mathematical framework that seeks to frame innovation in complex adaptive systems was also the focus of my doctoral work. This gave birth to an exciting period and will result in multiple series of books.

The first series, comprising of six books, extended my inquiry into mathematics and complex adaptive systems to an interesting limit culminating in the nature of Light and the Cosmos. The fractal mathematics I propose is at the heart of this series: Cosmology of Light.

The first book, 'A Story of Light: A Simple Exploration of the Creation and Dynamics of this Universe and Others' contains the main ideas in the mathematics, in non-mathematical terms, that are further explored mathematically in the remaining books in this series. The second book 'Oceans of Innovation: The Mathematical Heart of Complex Systems' describes my interpretation of the mathematical foundation of complex systems. The third book, 'Emergence: A Mathematical Journey from the Big Bang to Sustainable Global Civilization' applies the mathematics to several layers of matter and life. The fourth book, 'Quantum Certainty: A Mathematics of Natural and Sustainable Human History' describes a process culminating in space, time, energy, and gravity quantization by which history is made. The fifth book, 'Super-Matter: Functional Richness in an Expanding Universe' describes a process for the creation of super-matter-based on a need for continued functional-richness. A link is made between the resulting quantization of space and an expanding universe. The final book, 'Cosmology of Light: A Mathematical Integration of Matter, Life, History & Civilization, Universe, and Self' proposes an integrated mathematical framework that flows through all things, hence unifying matter, light, civilization, history, universe, and self.

The second series further explores the implications of "one mathematics flowing through all things". The first book in this series 'The Emperor's Quantum Computer: An Alternative Light-Centered Interpretation of Quanta, Superposition, Entanglement and the Computing that Arises from it' describes an alternative narrative for quantum computing to the one commonly expressed today. The second book in the series, 'The Origin and Possibilities of Genetics: A Mathematical Exploration in a Cosmology of Light' explores pre-genetic, genetic, and post-genetic possibilities in a cosmology of light. This book, 'The Second Singularity: A Mathematical Exploration of AI-Based and Other Singularities in a Cosmology of Light' explores the limits of AI-based singularities with respect to light-based singularities. This book, the final in this series explores transhumanism in a cosmology of light.

The third series, Artistic Interpretation of Cosmology of Light, is intended to make Cosmology of Light more accessible by interpreting it artistically. The first book in the series is 'The Mandala Illustrated Story of Light'. The objective is to lead the reader through the story of light using mandalas as an aid in the journey. Several additional books are envisioned and are currently under development.

About the Author

Dr. Pravir Malik has been developing a unified theory and mathematics of organization over the last three decades. He has written 16 books related to this to emphasize a whole systems approach integrating individual, organizational, economic, social, environmental, and evolutionary dimensions. In recent years he has been intimately involved with computer modeling of complex organizational, economic, and world systems to help different stakeholders practically navigate and understand possible futures. He is also a regular contributor to Forbes and recently completed a ten-part series on the creation of sustainable wealth through interpreting environments as complex adaptive systems. In 2020 he also designed and delivered a multi-part Organizational Sciences Certification program with Forbes that was attended by executives from 250 companies. This program included a path-breaking approach to bringing about organizational change by leveraging light and was derived from Dr. Malik's 10-book series on Cosmology of Light.

Dr. Malik has held a number of leadership positions. He is currently Chief Strategy Officer at Galaxies, focused on modeling of complex realities, and Chief Technologist at Deep Order Technologies where he is spearheading the development of a revolutionary atom-based quantum computer. Dr. Malik has deep interest in core technologies such as quantum computing, artificial intelligence, genetics, and transhumanism. His recent writings propose alternative trajectories of development for each of these areas to increase the likelihood of sustainable global development. Formerly he has been Head of Organizational Sciences at Zappos.com - an eCommerce company, Managing Director of BSR - a global sustainability and environmental consulting company, and a member of the Founding Team of A.T. Kearney India - a global operations-focused management consulting company.

He has a Ph.D. in Technology Management with a focus on Mathematics of Innovation in Complex Adaptive Systems from the University of Pretoria, an MBA from Northwestern University's J.L. Kellogg Graduate School of Management with a focus on Marketing and Organizational Behavior, an MS in Computer Science from the University of Florida with a focus on AI, and a BSE in Computer Engineering from the Case Western Reserve University. He has also served on the faculty of Sri Aurobindo International Center of Education at the undergraduate level. Pravir is a global citizen who has lived, worked, and been educated in many parts of the world.

About the Illustrator

Margaret Astrid Phanes illustrates illuminating graphics of visual meditation. She teaches Light-Force meditation which has inspired these works. Ms. Phanes taught digital graphics at the University of Hawaii Maui College and the University of California, Santa Cruz. Her visual meditations on Light - Force consciousness have been presented at conferences, in international publications, and online. Her portfolio may be viewed at www.margaretphanes.com.

Selected Author Online Presence

- Amazon Author Page: https://www.amazon.com/Pravir-Malik/e/B002JVAEZE
- LinkedIn Profile: https://www.linkedin.com/in/pravirmalik/
- Forbes Page & Articles: https://www.forbes.com/sites/forbeshumanresourcescouncil/people/pravirmalik1/#1fa1097c17be
- Forbes Profile: https://profiles.forbes.com/members/hr/profile/Pravir-Malik-Head-Organizational-Sciences-Zappos/44463250-f2ab-434a-b1e2-0a6bdf54d970
- Google Scholar Page: https://scholar.google.com/citations?user=7DWWWZ8AAAAJ&hl=en
- Sage Author Page: https://us.sagepub.com/en-us/nam/author/pravir-malik
- IEEE Profile: https://ieeexplore.ieee.org/author/37086022058

- YouTube Page: https://www.youtube.com/user/Aurosoorya
- Twitter: https://twitter.com/PravirMalik
- Research Gate Profile: https://www.researchgate.net/profile/Pravir_Malik
- Eventbrite: https://www.eventbrite.com/o/pravir-malik-30159112262
- Medium: https://medium.com/@PravirMalik
- Company website: http://www.deepordertechnologies.com/